100 Questions to Pass the PE

Practice Questions and Answers to Prepare for the Principles and Practice of Engineering Exam: HVAC and Refrigeration

By Steven Arms, PE

100 Questions to Pass the PE

ISBN 9781728630595

Copyright © 2018 by Steven Arms

No part of this publication may be reproduced, stored in a retrieval system or transmitted in any form or by any means, electronic, mechanical, photocopying, recording, scanning or otherwise, except as permitted under Sections 107 or 108 of the 1976 United States Copyright Act, without the prior written permission of the author. Requests to the author for permissions should be sent to 100questionshelp@gmail.com.

Errata may be sent to 100questionshelp@gmail.com

TABLE OF CONTENTS

Introduction……………………………………………………..…….………………………….1
Exam Part I – Principles…………………………………………..…….………………………3
Exam Part II – Applications……………………………………………..…….………………25
Answer Key……………………………………………………………..…….…………………49
Solutions Part I – Principles…………………………………………...…….………………...51
Solutions Part II – Applications……………………………………..…….…………………91
References…………………………………………………………….……….………………111

INTRODUCTION

Obtaining a Professional Engineering license marks a major milestone in an engineers' career, and comes with a number of benefits. These benefits include licensure to approve drawings and calculations, career advancement, and access to membership in professional groups. The amount of education and professional experience required to become a Professional Engineer varies from state to state, but in all cases applicants must pass the Principles and Practice of Engineering Examination to obtain licensure.

Preparation for the exam should begin months in advance. Because most candidates are also occupied with a career or schooling, a regular schedule of preparation is recommended. Topics to review include:

- Engineering units and conversions
- Engineering economics
- Basic electrical concepts
- Thermodynamics
- Fluid mechanics
- Heat transfer
- Psychrometrics
- HVAC systems
- Controls
- Air distribution
- Piping
- Refrigeration
- Air quality requirements
- Vibration
- Acoustics

Once a basic understanding of these concepts is achieved, a candidate should prepare for the exam with practice questions to improve test taking skills and to reveal common errors. The purpose of this book is to provide practice questions similar in difficulty to the exam questions and to provide clear solutions for review. It is written in two sections; Principles and Applications, and covers a breadth of topics that may be found on the exam. Although the exam itself is 80 questions, this book includes 105 questions for preparation.

Bringing all four ASHRAE Handbooks to the exam is highly recommended. ASHRAE Standards, academic textbooks, and professional handbooks can also be brought at the candidate's discretion. Because there is a time limit on the exam, all books that are brought to the exam should be reviewed thoroughly beforehand.

Candidates should arrive to the exam site early in case of traffic or other unexpected delays. Because exam sites typically host multiple engineering disciplines over a large geographical area, the administration of the test, such as reviewing identification cards, reading the exam rules, and handing out the exams, does take a significant amount of additional time. Even though the test itself comprised of 2 four hour long sessions, a candidate can expect the entire day to take about 11-12 hours.

It is critical to review the exam rules before the day of the exam so that a candidate can be properly prepared. Exam rules can be found online at the NCEES website.

EXAM PART I - PRINCIPLES

1. A heat pump unit consumes 5.0 kW of power and has a COP of 3.0. How much heat does it add to the space when the unit is in operation?

    A)   17,060 $\frac{Btu}{hr}$

    B)   34,120 $\frac{Btu}{hr}$

    C)   51,180 $\frac{Btu}{hr}$

    D)   68,240 $\frac{Btu}{hr}$

2. Two models of through-the-wall PTAC units are being compared for use in a hotel application. Both models provide 15,000 $\frac{Btu}{hr}$ of cooling, and there will be a total of 80 systems in the hotel. It is estimated that the units will run for 3 hours per day on average, and historically the cooling season lasts 85 days. The cost of electricity is $0.11/ kWh. Assuming the installation cost is the similar for both systems, what is the payback for system B?

    System A
    Initial Cost = $800 ea
    EER = 9.1

    System B
    Initial Cost = $1,100 ea
    EER = 12.2

    A)   7 years

    B)   8.2 years

    C)   11 years

    D)   25 years

3. A machine is purchased for $100,000 with a loan that requires a 20% down payment and has an interest rate of 6% APR. The lifetime of the loan is 60 months, and the salvage value of the machine at the end of 5 years is estimated to be $60,000. What is the monthly payment for the machine?

   A) $1,544
   B) $1,583
   C) $1,930
   D) $1,978

4. At 5,000 ft elevation, a room with SHR of 0.6 is cooled with 2,500 CFM of air supplied at 53 °F$_{db}$. The room is maintained at 77 °F$_{db}$ / 45% RH. How much moisture is expected to be removed from the air?

   A)   2.85 GPH
   B)   3.93 GPH
   C)   4.12 GPH
   D)   4.85 GPH

5. A room with a sensible cooling load of 120,000 $\frac{Btu}{hr}$ and latent cooling load of 120,000 $\frac{Btu}{hr}$ is cooled with supply air at 56 °F$_{db}$ / 30% RH. The room temperature is 78 °F$_{db}$. The wet bulb temperature of the room is most nearly:

   A)   50 °F$_{wb}$
   B)   61 °F$_{wb}$
   C)   64 °F$_{wb}$
   D)   67 °F$_{wb}$

6. Steam at 350 °F and atmospheric pressure is injected into 10,000 CFM of air at 65 °F$_{db}$ / 30% RH until the air is fully saturated. What is the dry-bulb temperature of the leaving air?

   A)   70 °F$_{db}$

   B)   73 °F$_{db}$

   C)   74 °F$_{db}$

   D)   76 °F$_{db}$

7. 2,500 CFM of outside air at 57 °F$_{db}$ / 20% RH ventilates a laboratory that is kept at 74 °F$_{db}$ / 45% RH. The laboratory has 16 permanent rotary evaporators that each evaporate 0.2 lb of water per hour. When the restaurant is unoccupied, what is the humidification load of the space?

   A) $62 \frac{lb_w}{hr}$

   B) $65.2 \frac{lb_w}{hr}$

   C) $68.4 \frac{lb_w}{hr}$

   D) $71.6 \frac{lb_w}{hr}$

8. 60 °F water is stored in an elevated storage tank that is vented to the atmosphere, as shown below. The water outlet pipe I.D. is 4". Neglecting friction losses, what is the height required to achieve 20 GPM of water flow at 80 psig at the end of the pipe?

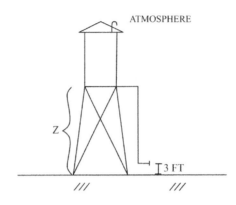

A)   151.1 ft

B)   154.3 ft

C)   188.1 ft

D)   204.2 ft

9. 8,000 CFM of air is heating from 65 °F to 115 °F as it passes through a coil, as shown below. Steam at 20 psia and 450 °F enters the coil at a rate of $14 \frac{lb_m}{min}$. Assuming the coil efficiency is 100%, what is the quality of the steam leaving the coil?

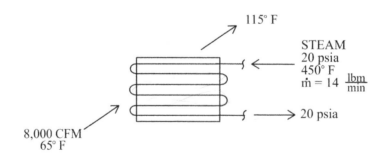

A)   0.00

B)   0.07

C)   0.48

D)   0.58

10. A Pitot-static tube is used to measure the velocity of air traveling through a duct. The Pitot tube manometer reading is 0.15" water column. If the project elevation is 3,000 ft and standard air temperature, what is the velocity of the air at the point measured?

   A)   1,560 FPM

   B)   1,640 FPM

   C)   1,960 FPM

   D)   2,350 FPM

11. A 13 BHP pump provides 30 GPM of water (SG =1.0) against 50 ft of head. If the head of the system increases to 55 ft, what will be the resulting BHP?

   A)   13.4 BHP

   B)   13.6 BHP

   C)   14.3 BHP

   D)   15.0 BHP

12. A 10 HP fan provides 3,000 CFM of air at a static pressure of 1.2" water column. What is the minimum motor size required (HP) in order for the fan to provide 6,000 CFM of air?

   A)   40

   B)   50

   C)   75

   D)   100

13. A condenser water system that includes a cooling tower, air separator, pump and heat exchanger is installed on a roof at an elevation of 3,000 ft. The induced draft cooling tower operates at 95 °F inlet water temperature and 85 °F outlet water temperature. The cooling tower is 12 ft tall, sits on 3 ft stands and has a basin which contains 1 ft of water. The pump is mounted on a skid so that the centerline of the pump is elevated 1 ft above the roof. If the friction losses in the pipe are 14.3 psig/ 1,000 ft, what is the NPSHA (ft) for the pump shown below?

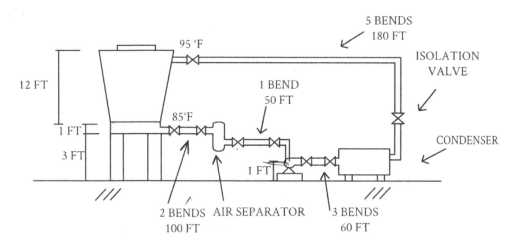

PRESSURE DROP AIR SEPARATOR = 3 FT   PRESSURE DROP ISOLATION VALVES = 0.5 FT/EA

PRESSURE DROP CONDENSOR = 20 FT   PRESSURE DROP PIPE BENDS = 2.5 FT/EA

A) 12.3

B) 14.6

C) 15.4

D) 17.4

14. In order to prevent freezing, an air conditioning unit will shut off if the evaporator coil temperature reaches below 34 °F. If the air entering the coil is at 75 °F$_{db}$/ 68 °F$_{wb}$ and the coil efficiency is 0.8, what is the minimum allowable LAT?

A) 40.8 °F

B) 42.2 °F

C) 42.6 °F

D) 48.8 °F

15. A steel plate with convective heat transfer coefficient of 8 $\frac{Btu}{°F \cdot ft^2 \cdot hr}$ has a surface temperature of 250 °F. If the emissivity of the plate is 0.75 and the ambient air temperature is 80 °F, what is the ratio of heat transfer by convection to heat transfer by radiation?

   A) 3.7

   B) 4.1

   C) 5.5

   D) 6.3

16. A pump is used to circulate 1.8 GPM of hot water at 140 °F (SG = 0.98) in a domestic water system that uses an electric water heater with a 50 gallon storage tank. The domestic water system has 1,200 ft of pipe with a friction loss of 2.0 psi/100 ft and has 80 ft of fitting losses. Heat is lost through the pipe at a rate of 0.6 $\frac{Btu/hr}{ft}$. The highest fixture is located 35 ft above the pump. The motor efficiency of the circulating pump is 88% and the pump efficiency is 78%. The efficiency of the electric water heater is 95%. If the cost of electricity is $0.12/ kWh and the pump operates continuously, what is the annual operating cost of the recirculating pump?

   A) $55

   B) $90

   C) $180

   D) $320

17. 1,000 CFM of outside air at 20 °F$_{db}$ / 20% RH is mixed with 2,000 CFM of return air at 85 °F$_{db}$ / 50% RH, then heated with a hot water coil with a coil temperature of 140 °F and coil efficiency of 58%. What is the supply air dry bulb leaving air temperature?

   A) 94.0 °F

   B) 95.5 °F

   C) 106.9 °F

   D) 107.8 °F

18. Water at 60 °F flows through a 6 foot long, 2 inch I.D. pipe at a mass flow rate of 17 $\frac{lb_m}{min}$. The Reynolds number of the water is most nearly:

   A) $3.17 \cdot 10^2$
   B) $3.22 \cdot 10^3$
   C) $1.33 \cdot 10^5$
   D) $1.33 \cdot 10^7$

19. 50 GPM of water flows through the piping circuit shown below. Including fitting losses, Path A has an equivalent length of 30 ft and Path B has an equivalent length of 160 ft. Pipe diameter remains constant throughout the system. Coil A has a pressure drop of 40 ft water column and Coil B has a pressure drop of 20 ft water column. What is the pressure drop required in Balance Valve A to achieve 30 GPM flow through Coil A?

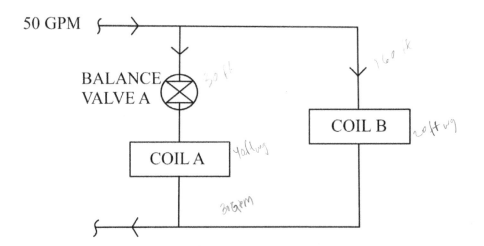

   A)   4.3 psig
   B)   17.4 psig
   C)   21.7 psig
   D)   38.7 psig

20. An AC unit removes 2.5 tons of heat from a space and has a compressor input of 2 HP. If the unit operates between 60 psig and 160 psig with R-22 as the working refrigerant, what is the required refrigerant mass flow rate? Assume refrigerant enters the compressor as a saturated vapor.

A) $61 \frac{lb_m}{hr}$

B) $70 \frac{lb_m}{hr}$

C) $360 \frac{lb_m}{hr}$

D) $410 \frac{lb_m}{hr}$

21. Air at 85 °F$_{db}$ / 40% RH enters a cooling coil and leaves at 64 °F$_{db}$ / 59 °F$_{wb}$. The coil apparatus dew point is most nearly:

A)   55 °F

B)   56 °F

C)   58 °F

D)   59 °F

22. An air cooled heat pump has a COP published by the manufacturer of 3.8. When the outside temperature is 47 °F and the LAT of the unit is 103 °F, what is the maximum theoretical COP?

A)   8.6

B)   9.1

C)   10.1

D)   15.5

23. A pump delivers water (SG = 1.0) against 30 ft of head in a 6" I.D. pipe. The average fluid velocity is 15 FPS. If the pump efficiency is 0.67 and the motor efficiency is 0.90, what is the input power required?

   A) 7.5 kW
   B) 11.2 kW
   C) 12.4 kW
   D) 49.6 kW

24. A 14" square supply plenum has 3 round supply takeoffs and a round bypass takeoff as shown below. What is the velocity of the bypass air?

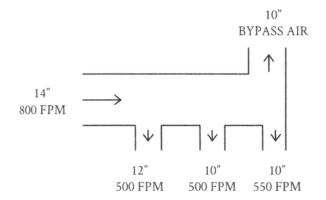

   A) 124 FPM
   B) 201 FPM
   C) 228 FPM
   D) 278 FPM

25. 2,560 CFM of air at 55 °F$_{db}$ / 51°F$_{wb}$ passes through a 10 kW electric heating coil. What is the relative humidity of the air after passing through the coil?

   A) 35%
   B) 40%
   C) 43%
   D) 50%

26. 3,000 CFM of air is heated and humidified by passing through an electric heating coil and steam injector as shown below. Air enters at 60 °F$_{db}$ / 20% RH. The heating coil consumes 44 kW of power, and the steam injector adds 22 GPH of water at 75 °F to the air. What is the relative humidity of the air leaving the steam injector?

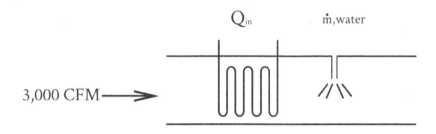

- A) 67%
- B) 75%
- C) 82%
- D) 85%

27. 1,400 CFM of air at 80 °F$_{db}$ / 67 °F$_{wb}$ is cooled and dehumidified by passing through a cooling coil that is below the entering air dew point temperature. The temperature of the water removed from the air is 60 °F. How much heat must be removed from the air so that the leaving air temperature is 55 °F$_{db}$ saturated?

- A) 40,800 $\frac{Btu}{hr}$
- B) 50,600 $\frac{Btu}{hr}$
- C) 56,000 $\frac{Btu}{hr}$
- D) 63,400 $\frac{Btu}{hr}$

28. 200 CFM of return air is recirculated through a fan coil from a room with a sensible heat load of 30,000 $\frac{Btu}{hr}$ and a total heat load of 42,600 $\frac{Btu}{hr}$. The room temperature is 75 °F$_{db}$ / 40% RH. The coil has a bypass factor of 0.2 and the coil temperature is 52 °F. How much moisture is removed from the air as it passes through the fan coil?

   A)   0.21 GPH

   B)   0.35 GPH

   C)   0.40 GPH

   D)   0.52 GPH

29. A boiler operates at 10 psig and produces 100 $\frac{lb_m}{hr}$ of steam. 20 $\frac{lb_m}{hr}$ of water returns to the boiler as a liquid, and 80 $\frac{lb_m}{hr}$ of water returns to the boiler as a gas. What is the entropy of the return mixture entering the boiler?

   A) $0.63 \frac{Btu}{lb_m \cdot °R}$

   B) $1.09 \frac{Btu}{lb_m \cdot °R}$

   C) $1.44 \frac{Btu}{lb_m \cdot °R}$

   D) $1.72 \frac{Btu}{lb_m \cdot °R}$

30. 5,000 CFM of air at 100 °F$_{db}$ / 20% RH is cooled by an evaporative cooler. How much water is absorbed by the air as it passes through the evaporative cooler if the air leaves at 80 °F$_{db}$? Assume the water has reached steady-state conditions.

   A)   10.5 GPH

   B)   11.5 GPH

   C)   12.4 GPH

   D)   13.5 GPH

31. Which of the following statements does not describe a Law of Thermodynamics?

    A) Entropy of an isolated system always increases.

    B) As the temperature of a system approaches absolute zero, entropy approaches a minimum value.

    C) The total amount of energy in an isolated system always decreases.

    D) If two systems are in thermal equilibrium with a third system, they are in thermal equilibrium with each other.

32. 2,000 CFM of return air at 75 °F$_{db}$ / 40% RH is mixed with 800 CFM of outside air at 90 °F$_{db}$ / 77 °F$_{wb}$ and passes through a cooling coil. The cooling coil bypass factor is 0.15, and the air leaves the coil at 55 °F$_{db}$ / 46 °F$_{wb}$. The sensible heat ratio of the room is most nearly:

    A) 0.55

    B) 0.60

    C) 0.70

    D) 0.85

33. 60 GPM of water at 60 °F flows through an 8" I.D. Schedule 40 steel pipe. A control valve is used to control flow through the pipe, as shown. If the control valve flow coefficient is 16, and the pressure gauge reads 55 psig, what is the pressure immediately downstream of the control valve?

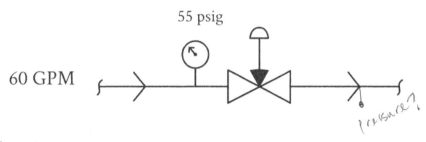

    A) 37 psig

    B) 41 psig

    C) 53 psig

    D) 55 psig

34. 160 °F water flows through a 2.9" I.D. pipe. The pipe O.D. is 3.5", and the thermal conductivity of the pipe is 20 $\frac{Btu}{hr \cdot °F \cdot ft}$. The temperature of the pipe outer surface is 158 °F. The pipe is wrapped with 2" insulation that has an R value of 5.0 $\frac{°F \cdot ft^2 \cdot hr}{Btu}$, and the temperature on the outside surface of the insulation is 82 °F. How much heat is lost in the pipe per linear foot?

A) 1,050 $\frac{Btu}{hr \cdot ft}$

B) 1,360 $\frac{Btu}{hr \cdot ft}$

C) 1,920 $\frac{Btu}{hr \cdot ft}$

D) 2,430 $\frac{Btu}{hr \cdot ft}$

35. Water flows through a 5" I.D. steel pipe at 15 FPM and is diverted into Path 1 and Path 2, as shown below. Path 1 has 5" I.D. pipe with friction factor f = 0.001 $\frac{in}{in}$, and Path 2 has 3" I.D. pipe with a friction factor f = 0.0008 $\frac{in}{in}$. If the equivalent length of Path 1 is 18 ft and the equivalent length of Path 2 is 36 ft, what is the flowrate through Path 2?

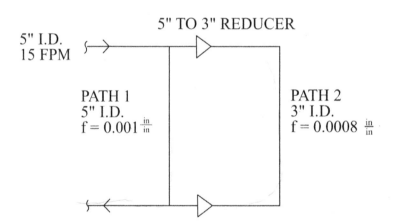

A) 2.5 GPM

B) 2.8 GPM

C) 11.6 GPM

D) 42.6 GPM

36. An air conditioning unit with a 5.0 kW compressor removes 27,000 $\frac{Btu}{hr}$ from a zone. What is the COP of the system?

   A)   1.6
   B)   1.8
   C)   3.0
   D)   3.6

37. The Net Refrigerant Effect is:

   A)   Equal to the compressor horsepower
   B)   Equal to the heat removed by refrigerant from the cooling medium
   C)   Equal to the heat rejected by the condenser
   D)   Equal to the heat absorbed by the evaporator plus the power consumed by the compressor

38. A steam boiler has an output of 116,000 $\frac{Btu}{hr}$. Steam leaves the boiler at 5 psig and 400 °F, and returns as a saturated liquid. If the system uses 2" O.D. pipe with wall thickness of ¼", what is the velocity of the steam as it leaves the boiler?

   A)   64 FPS
   B)   72 FPS
   C)   241 FPS
   D)   256 FPS

39. 90 GPM of water is pumped through 3" I.D. pipe, as shown below. If the isolation valve is suddenly closed, what is the maximum instantaneous pressure in the pipe?

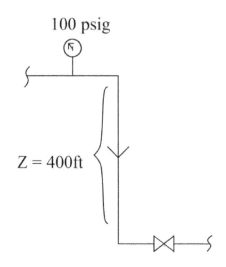

- A)   310 psig
- B)   445 psig
- C)   495 psig
- D)   530 psig

40. A refrigeration system using R-134a produces 10 tons of cooling. The compressor operates between 40 psia suction pressure and 300 psia discharge pressure. Assuming 60 °F superheat and no subcooling, what is the quality of the refrigerant as it leaves the expansion valve?

- A)   55%
- B)   60%
- C)   65%
- D)   80%

41. What is the COP of the refrigeration cycle shown below?

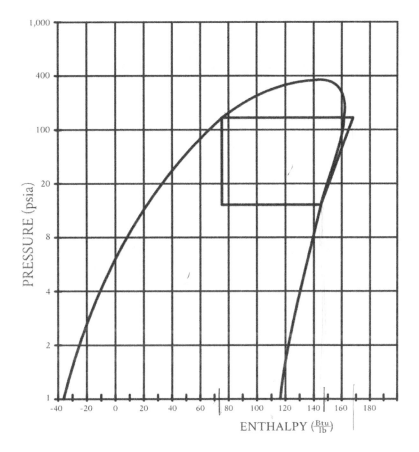

- A) 2.9
- B) 3.7
- C) 4.1
- D) 4.4

42. A sling thermometer is used to measure the outdoor air temperature and reads 72 °F$_{db}$ / 60 °F$_{wb}$. What is the partial pressure of water vapor in the air?

- A) 0.19 psia
- B) 0.23 psia
- C) 0.30 psia
- D) 0.39 psia

43. An air-water mixture is at 90 °F$_{db}$ / 42% RH. What is the dew point temperature of the mixture?

   A)   62 °F

   B)   64 °F

   C)   71 °F

   D)   73 °F

44. A fan that produces 2,500 CFM of air at standard air conditions operates at a location with an atmospheric pressure of 11.8 psia and ambient temperature of 85 °F. What is the actual air flow rate that the fan will produce under the non-standard air conditions?

   A)   1,650 CFM

   B)   1,950 CFM

   C)   3,200 CFM

   D)   3,780 CFM

45. R-22 is the working refrigerant in a vapor compression system. Refrigerant leaves the evaporator at 20 °F and is compressed to 300 psia. The compressor has an isentropic efficiency of 50% and the cooling effect is 70,000 Btu/hr. What is the power consumed by the compressor?

   A)   6.6 kW

   B)   13.2 kW

   C)   25.6 kW

   D)   45.0 kW

46. 8,000 CFM of air at 60 °F$_{db}$ / 30% RH is mixed with 15,000 CFM of air at 73 °F$_{db}$ / 50% RH. What is the wet bulb temperature of the resulting air mixture?

    A) 56 °F
    B) 63 °F
    C) 68 °F
    D) 69 °F

47. A counter flow heat exchanger uses 25 GPM of wastewater at 300 °F to preheat 60 GPM of fresh water at 80 °F. The heat transfer coefficient of the heat exchanger is 44 $\frac{Btu}{hr \cdot ft^2 \cdot °F}$, and a surface area of 148 ft². The temperature of the fresh water leaving the heat exchanger is most nearly:

    A) 115 °F
    B) 130 °F
    C) 155 °F
    D) 180 °F

48. The entropy of a liquid-vapor mixture of water at 280 °F is 0.762 $\frac{Btu}{lb_m \cdot °R}$. What is the average enthalpy of the mixture?

    A) 510 $\frac{Btu}{lb}$
    B) 544 $\frac{Btu}{lb}$
    C) 581 $\frac{Btu}{lb}$
    D) 653 $\frac{Btu}{lb}$

49. 15 $\frac{lb_m}{s}$ of steam at 300 °F, 5 psia enters a turbine, which produces 400 kW of power. If the steam leaves the turbine at 1 psia, what is the temperature of the steam exiting the turbine?

A) 240 °F
B) 252 °F
C) 274 °F
D) 287 °F

50. Air (k = 1.4) at 80 °F, 5 psig, is isentropically compressed to 20 psig. What is the temperature of the air after compression?

A) 175 °F
B) 490 °F
C) 765 °F
D) 950 °F

51. 2,000 CFM of air is heated and humidified from 40 °F$_{db}$ / 10% RH to 73 °F$_{db}$ / 45% RH. What is the latent heat load of the air?

A) 50,800 $\frac{Btu}{lb}$
B) 76,000 $\frac{Btu}{lb}$
C) 98,500 $\frac{Btu}{lb}$
D) 142,800 $\frac{Btu}{lb}$

Psychrometrics

52. If atmospheric pressure is measured to be 12.5 psia, and what is the partial pressure of water vapor in air at 92 °F$_{db}$ with a humidity ratio of 0.013 $\frac{lb_w}{lb_a}$?

- A) 0.19 psia
- B) 0.26 psia
- C) 0.29 psia
- D) 0.31 psia

EXAM PART II - APPLICATIONS

Heat transfer

53. What is the required R value of the 4" insulation in order to maintain an indoor air temperature of 72 °F if 8,000 $\frac{Btu}{hr}$ of heating is provided to the space? The area of the exposed wall is 2,000 ft². Assume still air conditions on both sides of the wall and neglect the effects of the studs.

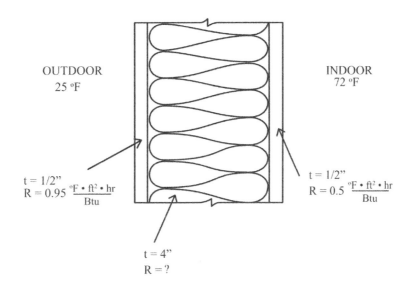

A) $7.6 \frac{°F \cdot ft^2 \cdot hr}{Btu}$

B) $8.4 \frac{°F \cdot ft^2 \cdot hr}{Btu}$

C) $8.9 \frac{°F \cdot ft^2 \cdot hr}{Btu}$

D) $10.3 \frac{°F \cdot ft^2 \cdot hr}{Btu}$

Chapter 9

54. An educational facility is adding a new building to its campus. The new building will have a 1,200 ft² art classroom that will seat 30 people and a 2,000 ft² computer laboratory that will seat 60 people. The minimum required ventilation rate for the new building is most nearly:

A) 1,260 CFM

B) 1,280 CFM

C) 1,360 CFM

D) 1,380 CFM

55. A beauty and nail salon is being built in a local strip mall. The area of the new salon is 800 ft² and is intended to have a maximum of 12 occupants. The minimum exhaust rate is:

   A)   450 CFM

   B)   480 CFM

   C)   500 CFM

   D)   520 CFM

Refrig

56. A 20 ton refrigeration unit is used to cool strawberries from 74 °F to 27 °F. How many pounds of strawberries can this system cool in a 24 hour period?

   A)   32,850 lbs

   B)   43,640 lbs

   C)   133,030 lbs

   D)   13,700 lbs

57. An ice maker produces 300 lbs of ice per hour and stores the ice at 14 °F. If the incoming water temperature is 70 °F, what is the minimum capacity of the ice making machine?

   A)   16,800 $\frac{Btu}{hr}$ Cp mass in book

   B)   36,000 $\frac{Btu}{hr}$

   C)   57,200 $\frac{Btu}{hr}$

   D)   79,200 $\frac{Btu}{hr}$

58. The rate of heat transfer per unit area of the wall shown below is most nearly:

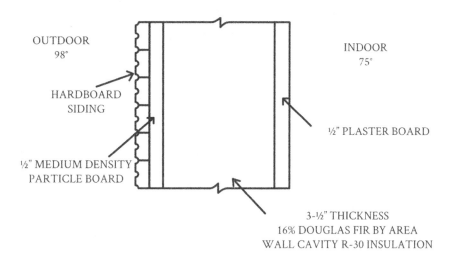

A) $0.83 \frac{\text{Btu/hr}}{\text{ft}^2}$

B) $0.89 \frac{\text{Btu/hr}}{\text{ft}^2}$

C) $1.74 \frac{\text{Btu/hr}}{\text{ft}^2}$

D) $2.34 \frac{\text{Btu/hr}}{\text{ft}^2}$

59. A 50 ft section of 4-½" OD Schedule 40 steel pipe is installed to distribute 300 °F steam throughout a building. The pipe was installed on a day with an ambient temperature of 60 °F. If the inside diameter of the pipe is 4", how far will the pipe expand when steam is introduced into the system?

A) 0.52 in

B) 0.93 in

C) 1.40 in

D) 1.76 in

60. A control damper that behaves as a first order, linear system is used to increase the airflow in a duct from 850 CFM to 1,100 CFM. The time constant of the system is 20 seconds. The damper is opened from 40% opened to 65% opened to achieve the increased flow. What is the gain of the process?

A) $7.9 \frac{CFM}{s}$

B) $12.5 \frac{CFM}{s}$

C) $10 \frac{CFM}{\%}$

D) $16.9 \frac{CFM}{\%}$

61. A programmable thermostat reads a space temperature of 62 °F. The thermostat is set to so that the system provides heating when the space temperature is below 72 °F and to provide cooling when the space temperature is above 78 °F. What is the thermostat deadband?

A) 10 °F

B) 6 °F

C) 3 °F

D) 2 °F

62. An exhaust fan has a rate of rotation of 1,200 rpm. The fan weighs 700 lb_m and sits on 4 rubber isolators with stiffness k = 400 $\frac{lb_f}{in}$. What is the natural frequency of the exhaust fan?

A) 0.68 Hz

B) 1.36 Hz

C) 2.36 Hz

D) 4.73 Hz

63. The single shell, single pass heat exchanger shown below has a surface area of 20 ft² and overall heat transfer coefficient of 1,200 $\frac{Btu}{°F \cdot ft^2 \cdot hr}$. What is the heat transfer, in tons, between the hot and cold fluid?

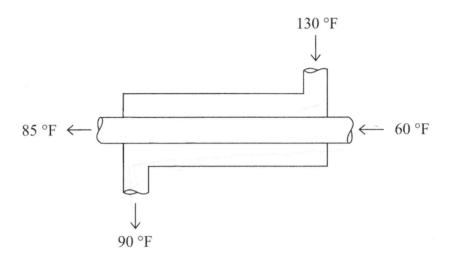

- A) 43 Tons
- B) 49 Tons
- C) 64 Tons
- D) 74 Tons

64. A closed feed water system is shown below. What is the temperature of the fluid at Point 4?

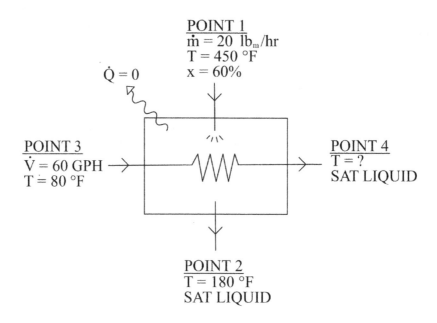

POINT 1
$\dot{m} = 20$ lb$_m$/hr
T = 450 °F
x = 60%

$\dot{Q} = 0$

POINT 3
$\dot{V} = 60$ GPH
T = 80 °F

POINT 4
T = ?
SAT LIQUID

POINT 2
T = 180 °F
SAT LIQUID

A) 110 °F
B) 118 °F
C) 122 °F
D) 328 °F

65. 2,500 GPM of water flows through 300 feet of 18" nominal Schedule 40 steel pipe without any bends, tees, reducers or valves. The friction factor of the pipe is 0.025. What is the friction loss, in feet of water column, of the steel pipe?

A) 0.26 ft
B) 0.77 ft
C) 1.07 ft
D) 1.44 ft

66. A 0° deflection high sidewall grille measuring 24" x 48" has a free area of 980 in². What is the throw of the outlet if the airflow rate is 2,000 CFM and the centerline velocity is 200 FPM?

A) 19 ft
B) 22 ft
C) 27 ft
D) 29 ft

67. A 12" by 30" duct delivers 8,000 CFM of air down a 125 ft corridor. What is the friction loss of the duct?

A) 0.75 in w.g.
B) 0.83 in w.g.
C) 0.88 in w.g.
D) 1.00 in w.g.

68. R-152a is used as a propellant in aerosol spray cans and as working refrigerant in automotive air conditioning systems. The refrigerant has a lower flammability propagation and lower toxicity potential. What safety class does R-152a belong to?

A) A1
B) A2
C) B2
D) B3

69. Turning vanes are being added to a duct elbow to decrease the fitting pressure. The duct is 40" by 30" and is designed for 6,000 CFM of supply air. If the turning vanes reduce the fitting loss coefficient from 0.32 to 0.22, what is the fitting pressure of the elbow?

    A)   0.007 in w.g.

    B)   0.039 in w.g.

    C)   0.071 in w.g.

    D)   0.494 in w.g.

70. The purpose of a liquid line solenoid valve in an AC unit is:

    A)   To control the amount of refrigerant released to the evaporator coil

    B)   To prevent a flooded compressor start

    C)   To change the direction of the refrigerant flow

    D)   To prevent migration of refrigerant to the condenser coil when the unit is off

71. What is the sensible effectiveness of the heat recovery ventilator shown below? Assume C_p remains constant as temperature changes.

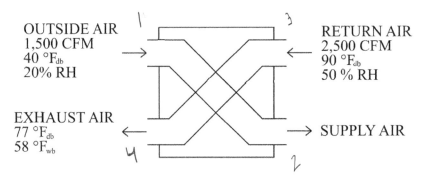

    A)   0.33

    B)   0.38

    C)   0.48

    D)   0.60

72. An industrial process which uses 100 GPM of ammonia as the working fluid is used to pre-heat 60 GPM of water. The initial water temperature is 60 °F. Assuming negligible heat losses in the heat exchanger, the leaving water temperature is most nearly:

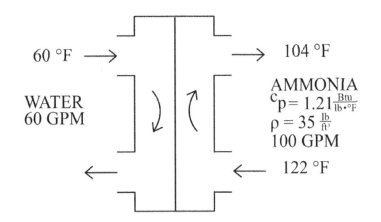

- A) 78 °F
- B) 80 °F
- C) 82 °F
- D) 97 °F

73. For a 1-1/2 inch nominal diameter pipe, which type of copper tube will have the smallest inside diameter?

- A) K
- B) L
- C) M
- D) DWV

74. A cooling tower has an efficiency of 0.65. 600 GPM of water enters the cooling tower at 95 °F, and the ambient air conditions are 80 °F$_{db}$ / 30% RH. What is the capacity of the cooling tower?

   A)   370 Tons

   B)   440 Tons

   C)   570 Tons

   D)   850 Tons

75. A cooling tower operates at 5,000 ft elevation and ambient air conditions of 75 °F$_{db}$ / 65 °F$_{wb}$. 15,000 CFM of air passes through the cooling tower and leaves at 75 °F saturated. What is the maximum allowable flow rate if the range required is 12 °F? Assume the approach of the cooling tower is negligible.

   A)   80.3 GPM

   B)   93.4 GPM

   C)   96.5 GPM

   D)   112.3 GPM

76. A furnace has an AFUE of 95%. Which of the following statements is false:

   A)   The furnace has a lower flue gas temperature than a similar non-condensing furnace

   B)   The furnace requires a metal flue vent to prevent melting

   C)   The furnace has a higher output capacity than a condensing furnace with the same input capacity

   D)   The furnace has two gas heat exchangers

77. A single pane window is shown below. Which equation best represents the rate of heat transfer through the window per unit area?

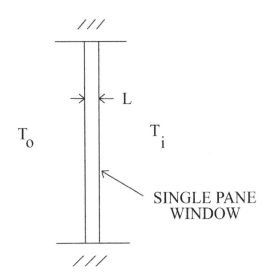

A) $\dfrac{1}{R_{outdoor\,air\,film} + R_{indoor\,air\,film} + R_{window}}$

B) $\dfrac{\Delta T}{\dfrac{1}{h_{outdoor\,air\,film}} + \dfrac{1}{h_{indoor\,air\,film}} + \dfrac{L}{k_{window}}}$

C) $\dfrac{\Delta T}{A\,(R_{outdoor\,air\,film} + R_{indoor\,air\,film} + L \cdot k_{window})}$

D) $\dfrac{\Delta T}{R_{window}}$

78. An air handling unit with a mass of 2,600 lb$_m$ is hung with 4 spring isolators each with a spring constant K of 830 $\dfrac{lb_f}{in}$. What is the deflection of each isolator?

A) 0.03"

B) 0.25"

C) 0.80"

D) 1.00"

79. A heat recovery system is shown below. The working fluid is chemical treated water, and has a specific heat of 0.87 $\frac{Btu}{lb \cdot F}$ and standard gravity of 1.07. Assume standard atmospheric pressure. What is the supply air leaving temperature?

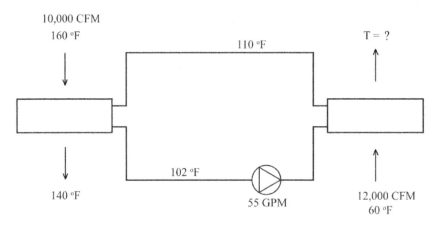

- A) 75.2 °F
- B) 75.8 °F
- C) 76.5 °F
- D) 77.0 °F

80. 4,000 CFM of air at 70 °F flows through a rectangular plenum with inlet duct dimensions of 16" by 20". What is the cutoff frequency (Hz) of the plenum?

- A) 246 Hz
- B) 327 Hz
- C) 338 Hz
- D) 409 Hz

81. Which of the following is not a purpose of a triple duty valve?

- A) Check valve
- B) Isolation valve
- C) Balance valve
- D) Mixing valve

82. The supply and return sides of an air distribution system have an equivalent length of 1,300 ft and 650 ft, respectively. The fan in the equipment selected can produce 4,800 CFM and the available external static pressure of the fan is 1.6" w.g. What is the friction rate per 100 ft that the air distribution system should be designed to?

A) 0.062

B) 0.077

C) 0.080

D) 0.082

83. What is the approach of the single pass heat exchanger shown below?

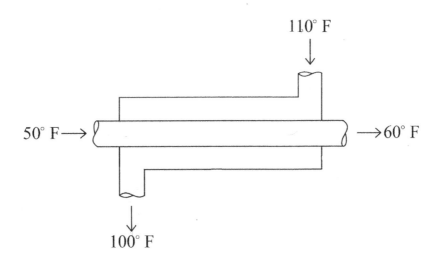

A) 10 °F

B) 40 °F

C) 50 °F

D) 60 °F

84. A fan operates at 6,000 CFM and 2 inches w.g. of static pressure in a duct that is 16" by 16". What is the mechanical efficiency of the fan at this operating point? Manufacturer fan data is supplied below.

	SP = 0.5 in wg		SP = 1.0 in wg		SP = 1.5 in wg	
CFM	n (rpm)	P (BHP)	n (rpm)	P (BHP)	n (rpm)	P (BHP)
4,000	287	0.15	315	0.43	315	0.79
4,500	323	0.22	355	0.62	355	1.13
5,000	359	0.30	394	0.85	394	1.55
5,500	394	0.40	433	1.12	433	2.07
6,000	430	0.52	473	1.46	473	2.68
6,500	466	0.66	512	1.86	630	3.41

	SP = 2.0 in wg		SP = 2.5 in wg		SP = 3.0 in wg	
CFM	n (rpm)	P (BHP)	n (rpm)	P (BHP)	n (rpm)	P (BHP)
4,000	440	1.22	459	1.71	505	3.14
4,500	495	1.74	516	2.43	568	4.47
5,000	550	2.39	573	3.34	631	6.14
5,500	605	3.18	630	4.45	694	8.17
6,000	660	4.13	688	5.77	757	10.60
6,500	715	5.25	745	7.34	820	13.48

A) 54%
B) 59%
C) 62%
D) 67%

85. An evaporative air cooler uses 2 GPM of water to temper 900 CFM of air. The water enters the cooler at 60 °F and leaves at 77 °F. The air enters the cooler at 80 °F$_{db}$ / 52 °F dew point, and leaves at 72 °F$_{db}$. What is the saturation efficiency of the evaporative air cooler?

A) 0.17
B) 0.40
C) 0.43
D) 0.64

86. A double seated globe valve:

    A)   Requires larger actuator force than a single seated globe valve

    B)   Is designed so that the pressure acting against the valve is balanced

    C)   Is used to increase the flow more rapidly through a valve

    D)   Is used where tight shutoff is required

87. A cooling tower operates at a flow rate of 550 GPM. 140,000 CFM of air enters the cooling tower at 90 °F_{db} / 30% RH, and the air leaves the cooling tower at 95 °F_{db} / 60% RH. What is the blowdown required if the cycles of concentration is kept below 5? Assume drift losses are 0.2%.

    A)   2.6 GPM

    B)   5.0 GPM

    C)   12.1 GPM

    D)   16.8 GPM

88. An exterior wall is built as shown below. The area of the wall is 1,000 ft², and 10% of the wall is uninsulated due to framing. Assuming 15 MPH winds outdoors, what is the average R value of the wall?

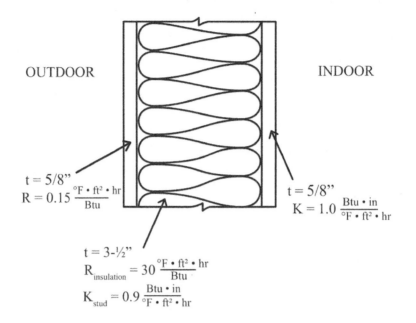

- A) 26.7
- B) 28.3
- C) 29.0
- D) 33.0

89. A boiler consumes 135 kW of energy and produces 400 $\frac{lb_m}{hr}$ of steam at 20 psig, 300 °F. If the feedwater temperature is 200 °F, what is the efficiency of the boiler?

- A) 80%
- B) 85%
- C) 87%
- D) 89%

90. A variable refrigerant flow system using R-410a is installed in a building with three zones. Information on the three zones is shown below. What is the maximum refrigeration charge allowed in the system?

Zone A
Volume = 6,600 ft^3
Max Occupancy = 220 people

Zone B
Volume = 4,800 ft^3
Max Occupancy = 150 people

Zone C
Volume = 6,000 ft^3
Max Occupancy = 110 people

- A) 80 lbs
- B) 105 lbs
- C) 115 lbs
- D) 124 lbs

91. Which of the following inspection techniques is used to detect corrosion, wall thinning, and cracking in the evaporator and condenser tubes of a chiller?

- A) Magnetic Particle Testing
- B) Eddy Current Testing
- C) Dye Penetrant Testing
- D) Ultrasonic Testing

92. The window shown below has a U value of 1.5 $\frac{Btu}{hr \cdot ft^2 \cdot °F}$ and a SHGC of 0.8. If the direct beam radiation is measured to be 1,400 $\frac{Btu}{hr \cdot ft^2}$, what is the solar heat gain through the window?

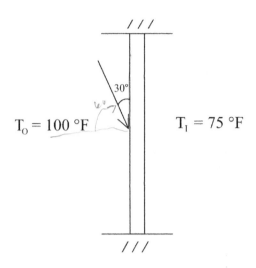

WINDOW AREA = 120 ft²

T_o = 100 °F T_I = 75 °F

- A) 67,200 $\frac{Btu}{hr}$
- B) 116,400 $\frac{Btu}{hr}$
- C) 134,400 $\frac{Btu}{hr}$
- D) 168,000 $\frac{Btu}{hr}$

93. Which society/ association publishes standards for duct and pipe insulation?

- A) ASTM
- B) AMCA
- C) AHRI
- D) CSA

94. In a constant volume system with supply air temperature reset, as the return air temperature decreases, the supply air temperature should:

    A) Increase

    B) Decrease

    C) Stay constant

    D) Reset to maximum value

95. A VAV terminal box serves a room that is 4,000 ft³ in volume and will be mounted 18 ft away from the nearest occupant. The sound power levels have been measured across the octave band and are shown in the table below. What is the Noise Criteria rating for this room?

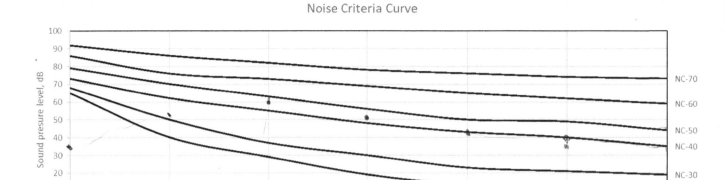

Frequency (Hz)	Sound Power Level (db)
125	35
250	51
500	60
1,000	62
2,000	52
4,000	42
8,000	45

A) NC-60

B) NC-50

C) NC-40

D) NC-30

96. An energy recovery ventilator has a fixed plate heat exchanger with a sensible effectiveness of 0.75 and a pressure drop of 1.0 inches water column. 500 CFM of exhaust air at 70 °F and 600 CFM of supply air at 30 °F passes through the ERV. If the fan efficiency is 0.7, the total fan power required is most nearly:

A)   0.11 HP

B)   0.13 HP

C)   0.25 HP

D)   0.37 HP

97. 4,000 CFM of outdoor air at 45 °F$_{db}$ / 10% RH and 4,000 CFM of exhaust air at 85 °F$_{db}$ / 30% RH enters a heat pipe with sensible effectiveness of 0.52. What is the sensible energy recovered from the heat pipe?

A)   64,600 $\frac{Btu}{hr}$

B)   86,200 $\frac{Btu}{hr}$

C)   97,000 $\frac{Btu}{hr}$

D)   105,500 $\frac{Btu}{hr}$

98. Fan performance curves are shown below for a fan tested over range of operating RPMs. What is the BHP required for the fan to operate at 6,000 CFM?

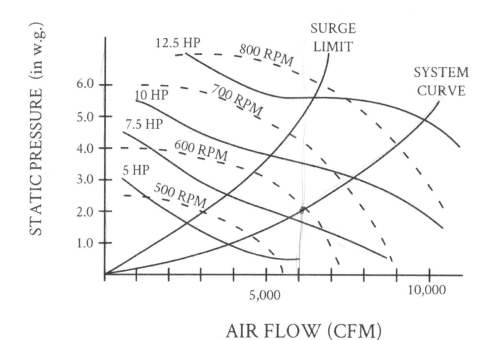

- A) 5 HP
- B) 7.5 HP
- C) 10 HP
- D) 12.5 HP

99. What is the thermal conductivity of exfoliated vermiculite insulation with a density of $5.5 \frac{lb}{ft^3}$?

- A) $3.2 \frac{Btu \cdot in}{hr \cdot ft^3 \cdot °F}$
- B) $3.9 \frac{Btu \cdot in}{hr \cdot ft^3 \cdot °F}$
- C) $4.4 \frac{Btu \cdot in}{hr \cdot ft^3 \cdot °F}$
- D) $4.7 \frac{Btu \cdot in}{hr \cdot ft^3 \cdot °F}$

100. A laboratory that is kept at 75 °F requires 12 air changes per hour for proper ventilation. The volume of the conference room is 9,200 ft³. If the outside air temperature is 35 °F, what is the heating load created by ventilation air?

A)   14,500 $\frac{\text{Btu}}{\text{hr}}$

B)   32,000 $\frac{\text{Btu}}{\text{hr}}$

C)   56,400 $\frac{\text{Btu}}{\text{hr}}$

D)   80,000 $\frac{\text{Btu}}{\text{hr}}$

ANSWER KEY

Question	Answer
1.	C
2.	D
3.	A
4.	C
5.	B
6.	C
7.	B
8.	C
9.	D
10.	B
11.	D
12.	D
13.	B
14.	B
15.	D
16.	D
17.	C
18.	B
19.	A
20.	D
21.	A
22.	C
23.	C
24.	C
25.	D
26.	A
27.	B
28.	A
29	C
30.	B
31.	C
32.	B
33.	B
34.	B

Question	Answer
35.	B
36.	A
37.	B
38.	A
39.	D
40.	A
41.	C
42.	A
43.	B
44.	C
45.	B
46.	A
47.	A
48.	A
49.	A
50.	A
51.	B
52.	B
53.	C
54.	C
55.	B
56.	A
57.	C
58.	A
59.	C
60.	C
61.	B
62.	D
63.	B
64.	A
65.	C
66.	B
67.	C

Question	Answer
68.	B
69.	A
70.	B
71.	B
72.	B
73.	A
74.	C
75.	B
76.	B
77.	B
78.	C
79.	B
80.	C
81.	D
82.	D
83.	C
84.	C
85.	D
86.	B
87.	A
88.	C
89.	D
90.	D
91.	B
92.	A
93.	A
94.	A
95.	C
96.	C
97.	B
98.	C
99.	C
100.	D

SOLUTIONS PART I - PRINCIPLES

1. $\text{COP}_{HP} = \frac{Q_{IN} + W_{IN}}{W_{IN}}$

   $W_{IN} = 5.0 \text{ kW} \cdot 3412.1 \frac{\text{Btu/hr}}{\text{kW}}$

   $W_{IN} = 17{,}060.5 \frac{\text{Btu}}{\text{hr}}$

   $Q_{IN} = (\text{COP}_{HP} \cdot W_{IN}) - W_{IN}$

   $Q_{IN} = (3.0 \cdot 17{,}060.5 \frac{\text{Btu}}{\text{hr}}) - 17{,}060.5 \frac{\text{Btu}}{\text{hr}}$

   $Q_{IN} = 34{,}121 \frac{\text{Btu}}{\text{hr}}$

   $Q_{OUT} = Q_{IN} + W_{IN}$

   $Q_{OUT} = 34{,}121 \frac{\text{Btu}}{\text{hr}} + 17{,}060.5 \frac{\text{Btu}}{\text{hr}}$

   $Q_{OUT} = 51{,}181.5 \frac{\text{Btu}}{\text{hr}}$

   The correct answer is C.

2. $\text{EER} = \frac{Q_{in} \,(\text{Btu}/\text{hr})}{W_{in} \,(\text{Watts})}$

   For system A,

   $W_{in} = \frac{15{,}000 \; \text{Btu}/\text{hr}}{9.1 \; \text{EER}}$

   $W_{in} = 1{,}648 \text{ W} \cdot \frac{1 \text{ kW}}{1{,}000 \text{ W}}$

   $W_{in} = 1.648 \text{ kW}$

   For system B,

   $W_{in} = \frac{15{,}000 \; \text{Btu}/\text{hr}}{12.2 \; \text{EER}}$

   $W_{in} = 1{,}229 \text{ W} \cdot \frac{1 \text{ kW}}{1{,}000 \text{ W}}$

   $W_{in} \text{ watts} = 1.229 \text{ kW}$

   Using simple payback method,

   Initial Cost$_A$ + Rate Consumption$_A$ · Time = Initial Cost$_B$ + Rate Consumption$_B$ · Time

$800 + 1.648 \text{ kW} \cdot \frac{3 \text{ hr}}{\text{day}} \cdot \frac{85 \text{ day}}{\text{year}} \cdot \frac{\$0.11}{\text{kWh}} \cdot T = \$1,200 + 1.229 \text{ kW} \cdot \frac{3 \text{ hr}}{\text{day}} \cdot \frac{85 \text{ day}}{\text{year}} \cdot \frac{\$0.11}{\text{kWh}} \cdot T$

$T = 25$ years

The correct answer is D.

3. 20% down payment reduces the loan to $80,000. The $60,000 salvage value is only realized if the machine is sold, after the loan is paid off.

Interest rate compounded monthly $= \frac{6\% \text{ APR}}{12 \text{ months}}$

$i = 0.5\%$

$(A/P, 0.5\%, 60) = \$80,000 \cdot (0.0193)$

$(A/P, 0.5\%, 60) = \$1,544$

The correct answer is A.

4. From ASHRAE Psychrometric Chart #4,

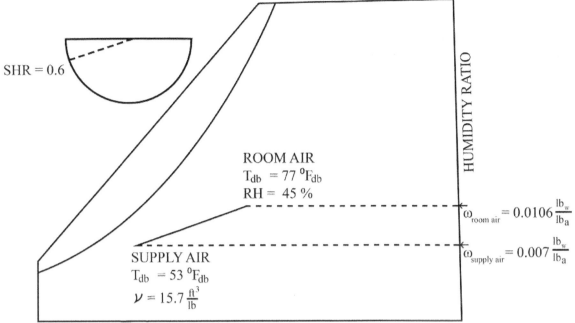

$\omega_{\text{supply air}} = 0.007 \frac{lb_w}{lb_a}$

$\omega_{\text{room air}} = 0.0106 \frac{lb_w}{lb_a}$

$\dot{m}_w = V_{\text{air}} \cdot \rho_{\text{air}} \cdot \Delta\omega$

$\dot{m}_w = 2{,}500 \frac{ft^3}{m} \cdot \frac{1\, lb_a}{15.7\, ft^3} \cdot (0.0106 \frac{lb_w}{lb_a} - 0.007 \frac{lb_w}{lb_a})$

$\dot{m}_w = 0.57 \frac{lb_w}{m}$

$\dot{m}_w = 0.57 \frac{lb_w}{m} \cdot 60 \frac{m}{hr} \cdot \frac{1\, ft^3}{62.4\, lb_m} \cdot \frac{7.48\, gal}{1\, ft^3}$

$\dot{m}_w = 4.12$ GPH

The correct answer is C.

5. $SHR = \dfrac{Q_S}{Q_S + Q_L}$

$SHR = \dfrac{120{,}000\,\frac{Btu}{hr}}{120{,}000\,\frac{Btu}{hr} + 120{,}000\,\frac{Btu}{hr}}$

$SHR = 0.5$

From ASHRAE Psychrometric Chart #1, $T_{room,\,wb} = 61.3\ °F$

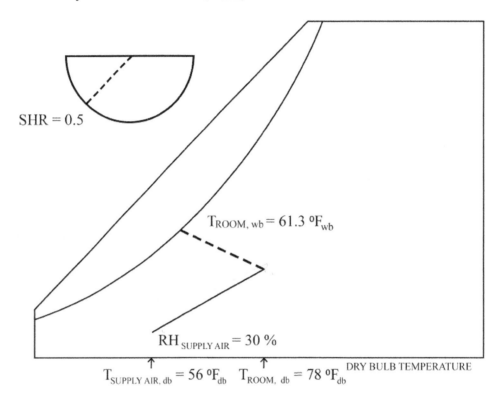

The correct answer is B.

6. From superheated steam tables for p = 14.7 psia,

$h_{steam,\,300°F} = 1{,}193.8\,\dfrac{Btu}{lb}$

$h_{steam,\,400°F} = 1{,}240.6\,\dfrac{Btu}{lb}$

$h_{steam,\,350°F} = 1{,}193.8\,\dfrac{Btu}{lb} + \dfrac{1{,}240.6\,Btu/lb - 1{,}193.8\,Btu/lb}{2}$

$h_{steam,\,350°F} = 1{,}217.2\,\dfrac{Btu}{lb}$

From ASHRAE Psychrometric Chart #1, $T_{db} = 74\ °F$

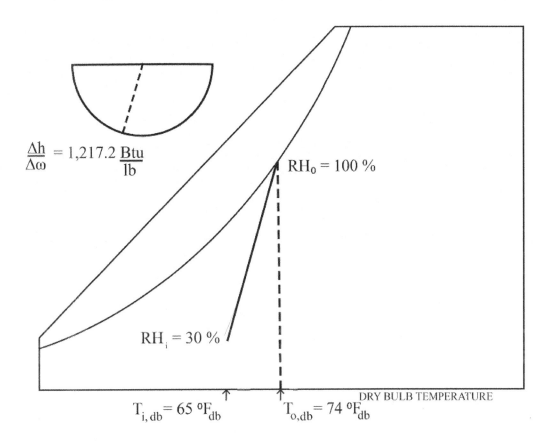

The correct answer is C.

7. From ASHRAE Psychrometric Chart #1,

$\omega_o = 0.002\ \frac{lb_w}{lb_a}$

$\omega_i = 0.008\ \frac{lb_w}{lb_a}$

$H = \rho \cdot V \cdot \Delta\omega$

$H_{ventilation} = 0.076\ \frac{lb_a}{ft^3} \cdot 2{,}500\ \frac{ft^3}{min} \cdot (0.008\ \frac{lb_w}{lb_a} - 0.002\ \frac{lb_w}{lb_a}) \cdot \frac{60\ min}{hr}$

$H_{ventilation} = 68.4\ \frac{lb_w}{hr}$

$H_{warmers} = 16 \cdot 0.2\ \frac{lb_w}{hr}$

$$H_{warmers} = 3.2 \frac{lb_w}{hr}$$

$$H_{Total} = H_{ventilation} - H_{warmers}$$

$$H_{Total} = 68.4 \frac{lb_w}{hr} - 3.2 \frac{lb_w}{hr}$$

$$H_{Total} = 65.2 \frac{lb_w}{hr}$$

The correct answer is B.

8. Solve for velocity of water at pipe outlet

$$A = \pi \cdot (0.166 \text{ ft})^2$$

$$A = 0.0873 \text{ ft}^2$$

$$Q = \frac{20 \text{ gal}}{\text{min}} \cdot \frac{1 \text{ ft}^3}{7.48 \text{ gal}} \cdot \frac{\cancel{62.4 \text{ lb}}}{\cancel{\text{ft}^3}} \cdot \frac{1 \text{ min}}{60 \text{ s}}$$

$$Q = 2.78 \frac{\text{ft}^3}{\text{s}}$$

$$V = \frac{Q}{A}$$

$$V = \frac{2.78 \text{ ft}^3/\text{s}}{0.0873 \text{ ft}^2}$$

$$V = 31.85 \frac{\text{ft}}{\text{s}}$$

Convert pressure required at pipe outlet from psig to psf

$$P_2 = 80 \frac{lb_f}{in^2} \cdot 144 \frac{in^2}{ft^2}$$

$$P_2 = 11{,}520 \frac{lb_f}{ft^2}$$

From Bernoulli equation,

$$z_1 = \frac{P_2}{\gamma} + \frac{v^2}{2g} + z_2$$

$$z_1 = \frac{11{,}520 \, lb_f/ft^2}{62.4 \, lb_f/ft^3} + \frac{(31.85 \, ft/s)^2}{2 \cdot (32.2 \, ft/s^2)} + 3 \text{ ft}$$

$$z_1 = 188.1 \text{ ft}$$

The correct answer is C.

9. $Q_{air} = 1.08 \cdot CFM \cdot \Delta T$

   $Q_{air} = 1.08 \cdot 8{,}000 \text{ CFM} \cdot (115\,°F - 65\,°F)$

   $Q_{air} = 432{,}000 \dfrac{Btu}{hr}$

   From superheated steam tables,

   $h_{400°F,\,20\,psia} = 1{,}239.3 \dfrac{Btu}{lb}$

   $h_{500°F,\,20\,psia} = 1{,}286.9 \dfrac{Btu}{lb}$

   $h_{450°F,\,20\,psia} = \dfrac{1{,}286.9\,Btu/lb + 1{,}239.3\,Btu/lb}{2}$

   $h_{450°F,\,20\,psia} = 1{,}263.1 \dfrac{Btu}{lb}$

   $Q_{steam} = \dot{m} \cdot (h_1 - h_2)$

   $h_2 = h_1 - \left(\dfrac{Q}{\dot{m}}\right)$

   $h_2 = 1{,}263.1 \dfrac{Btu}{lb} - \left(\dfrac{432{,}000\,Btu/hr}{14\,lb_m/min \cdot 60\,min/hr}\right)$

   $h_2 = 748.8 \dfrac{Btu}{lb}$

   $x = \dfrac{h_{mix} - h_f}{h_{fg}}$

   From saturated steam tables for p = 20 psia,

   $h_f = 196.3 \dfrac{Btu}{lb}$

   $h_{fg} = 959.9 \dfrac{Btu}{lb}$

   $x = \dfrac{748.8\,Btu/lb - 196.3\,Btu/lb}{959.9\,Btu/lb}$

   $x = 0.575$

   The correct answer is D.

10. From ASHRAE Fundamentals, Chapter 1,

$P_{3,000 \text{ ft}} = 13.17$ psia

$\rho_{air} = 0.075 \frac{lb_m}{ft^3} \cdot \frac{13.17 \text{ psia}}{14.7 \text{ psia}}$

$\rho_{air} = 0.067 \frac{lb_m}{ft^3}$

$V = \sqrt{2 \cdot g \cdot h}$

$V = \sqrt{2 \cdot 32.2 \frac{ft}{s^2} \cdot 0.15 \text{ in} \cdot \frac{1 \text{ ft}}{12 \text{ in}} \cdot \frac{62.4 \, lb_m/ft^3}{0.067 \, lb_m/ft^3} \cdot 60 \frac{\sec}{\min}}$

$V = 1{,}642.8$ FPM

The correct answer is B.

11. $\dfrac{P_2}{P_1} = \left(\dfrac{D_2}{D_1}\right)^3$

$\dfrac{h_2}{h_1} = \left(\dfrac{D_2}{D_1}\right)^2$

Combining equations gives

$\sqrt[3]{\dfrac{P_2}{P_1}} = \sqrt{\dfrac{h_2}{h_1}}$

$P_2 = P_1 \cdot \left(\dfrac{h_2}{h_1}\right)^{\frac{3}{2}}$

$P_2 = 13 \text{ HP} \cdot \left(\dfrac{55 \text{ ft}}{50 \text{ ft}}\right)^{\frac{3}{2}}$

$P_2 = 15$ BHP

The correct answer is D.

12. $\frac{HP_2}{HP_1} = \left(\frac{Q_2}{Q_1}\right)^2$

$HP_2 = 10 \text{ HP} \cdot \left(\frac{6,000 \text{ CFM}}{3,000 \text{ CFM}}\right)^2$

$HP_2 = 80 \text{ HP}$

The correct answer is D.

13. From ASHRAE Fundamentals, Chapter 1,

$P_{atm} = 13.17 \text{ psia}$

$h_{atm} = \frac{P_{atm}}{\gamma}$

$h_{atm} = \frac{13.17 \, ^{lb_f}/_{in^2} \cdot 144 \, ^{in^2}/_{ft^2}}{62.4 \, ^{lb_f}/_{in^2}}$

$h_{atm} = 30.4 \text{ ft}$

Averaging from saturated steam tables,

$h_{v.p.} = \frac{0.5778 \text{ psia} + 0.616 \text{ psia}}{2} \cdot \frac{2.3 \text{ ft water column}}{\text{psia}}$

$h_{v.p.} = 1.37 \text{ ft}$

$h_z = 3 \text{ ft} + 1 \text{ ft} - 1 \text{ ft}$

$h_z = 3 \text{ ft}$

$h_{AS} = 3 \text{ ft}$

$h_{iso \text{ valves}} = 4 \cdot 0.5 \text{ ft}$

$h_{iso \text{ valves}} = 2 \text{ ft}$

$h_{bends} = 3 \cdot 2.5 \text{ ft}$

$h_{bends} = 7.5 \text{ ft}$

$h_{pipe \text{ friction}} = 150 \text{ ft} \cdot \frac{14.3 \text{ psi}}{1,000 \text{ ft}}$

$$h_{\text{pipe friction}} = 2.145 \text{ psi} \cdot \frac{2.3 \text{ ft water column}}{\text{psi}}$$

$$h_{\text{pipe friction}} = 4.9 \text{ ft}$$

$$\text{NPSHA} = h_{\text{atm}} + h_z - h_{\text{v.p.}} - h_{\text{AS}} - h_{\text{iso valves}} - h_{\text{bends}} - h_{\text{pipe friction}}$$

$$\text{NPSHA} = 30.4 \text{ ft} + 3 \text{ ft} - 1.4 \text{ ft} - 3 \text{ ft} - 2 \text{ ft} - 7.5 \text{ ft} - 4.9 \text{ ft}$$

$$\text{NPSHA} = 14.6 \text{ ft}$$

The correct answer is B.

14. $BF = 1 - \eta$

    $BF = 1 - 0.8$

    $BF = 0.2$

    $$BF = \frac{T_{db,out} - T_{coil}}{T_{db,in} - T_{coil}}$$

    $$T_{db,out} = BF \cdot (T_{db,in} - T_{coil}) + T_{coil}$$

    $$T_{db,out} = 0.2 \cdot (75 \text{ °F} - 34 \text{ °F}) + 34 \text{ °F}$$

    $$T_{db,out} = 42.2 \text{ °F}$$

    The correct answer is B.

15. $\frac{Q_{conv}}{A} = h \cdot (T_{surface} - T_{ambient})$

    $\frac{Q_{conv}}{A} = 8 \frac{\text{Btu}}{\text{°F} \cdot \text{ft}^2 \cdot \text{hr}} \cdot (250 \text{ °F} - 80 \text{ °F})$

    $\frac{Q_{conv}}{A} = 1{,}360 \frac{\text{Btu}}{\text{hr}}$

    $\frac{Q_{rad}}{A} = \sigma \cdot \varepsilon \cdot [(T_{surface} + 460)^4 - (T_{ambient} + 460)^4]$

    $\frac{Q_{rad}}{A} = 0.1713 \times 10^{-8} \frac{\text{Btu}}{\text{R} \cdot \text{ft}^2 \cdot \text{hr}} \cdot 0.75 \cdot [(250 \text{ °F} + 460)^4 - (80 \text{ °F} + 460)^4]$

$$\frac{Q_{rad}}{A} = 217 \frac{Btu}{hr}$$

$$\frac{Q_{conv}}{Q_{rad}} = \frac{1{,}360 \, Btu/hr}{217 \, Btu/hr}$$

$$\frac{Q_{conv}}{Q_{rad}} = 6.26$$

The correct answer is D.

16. The recirculating pump requires energy to run the pump, and energy to reheat the water due to heat lost in the pipe.

$$h_{total} = h_{pipe} + h_{fittings} + h_z$$

$$h_{total} = \left(1{,}200 \, ft \cdot \frac{2.0 \, psi}{100 \, ft} \cdot 2.3 \frac{ft \, water}{psi}\right) + 80 \, ft + 35 \, ft$$

$$h_{total} = 170.2 \, ft$$

$$P_{pump, \, hp} = \frac{h \cdot Q \cdot SG}{3956 \cdot \eta}$$

$$P_{pump, \, hp} = \frac{170.2 \, ft \cdot 1.8 \, GPM \cdot 0.98}{3956 \cdot 0.88 \cdot 0.78}$$

$$P_{pump, \, hp} = 0.111 \, HP$$

$$P_{pump, \, kW} = 0.111 \, HP \cdot 0.7457 \frac{kW}{HP}$$

$$P_{pump, \, kW} = 0.083 \, kW$$

$$P_{heating} = \frac{Q}{\eta}$$

$$P_{heating} = \frac{0.6 \frac{Btu/hr}{ft} \cdot 1{,}200 \, ft}{0.95}$$

$$P_{heating} = 757 \frac{Btu}{hr} \cdot 0.000293 \frac{kW}{Btu/hr}$$

$$P_{heating} = 0.222 \, kW$$

$$P_{total} = P_{pump} + P_{heating}$$

$P_{total} = 0.083 \text{ kW} + 0.222 \text{ kW}$

$P_{total} = 0.305 \text{ kW}$

Operating cost = P · t · cost electricity

Operating cost = $0.305 \text{ kW} \cdot 24 \frac{\text{hr}}{\text{day}} \cdot 365 \frac{\text{day}}{\text{year}} \cdot \frac{\$0.12}{\text{kWh}}$

Operating cost = $320.62

The correct answer is D.

17. From ASHRAE Psychrometric Charts #1 and #2,

$v_{outside\ air} = 12.1 \frac{\text{ft}^3}{\text{lb}}$

$v_{return\ air} = 14.0 \frac{\text{ft}^3}{\text{lb}}$

$\dot{m}_{outside\ air} = \frac{1{,}000 \text{ ft}^3/\text{min}}{12.1 \text{ ft}^3/\text{lb}}$

$\dot{m}_{outside\ air} = 82.6 \frac{\text{lb}}{\text{min}}$

$\dot{m}_{return\ air} = \frac{2{,}000 \text{ ft}^3/\text{min}}{14.0 \text{ ft}^3/\text{lb}}$

$\dot{m}_{return\ air} = 142.8 \frac{\text{lb}}{\text{min}}$

$T_{mix} = T_a + \left(\frac{\dot{m}_b}{\dot{m}_a + \dot{m}_b}\right) \cdot (T_b - T_a)$

$T_{mix} = 20 \text{ °F} + \left(\frac{142.8 \text{ lb}/\text{min}}{82.6 \text{ lb}/\text{min} + 142.8 \text{ lb}/\text{min}}\right) \cdot (85 \text{ °F} - 20 \text{ °F})$

$T_{mix} = 61.2 \text{ °F}$

$BF = 1 - \eta$

$BF = 1 - 0.58$

$BF = 0.42$

$$BF = \frac{T_{db,out} - T_{coil}}{T_{db,in} - T_{coil}}$$

$$T_{db,out} = BF \cdot (T_{db,in} - T_{coil}) + T_{coil}$$

$$T_{db,out} = 0.42 \cdot (61.2\,°F - 140\,°F) + 140\,°F$$

$$T_{db,out} = 106.9\,°F$$

The correct answer is C.

18. $D = 2\,in \cdot \frac{1\,ft}{12\,in}$

    $D = 0.166\,ft$

    $A = \pi \cdot \left(\frac{0.166\,ft}{2}\right)^2$

    $A = 0.0216\,ft^2$

    $Q = \frac{\dot{m}}{\rho}$

    $Q = \frac{17\,lb_m/min}{62.4\,lb_m/ft^3} \cdot \frac{1\,min}{60\,sec}$

    $Q = 0.00454\,\frac{ft^3}{s}$

    $V = \frac{Q}{A}$

    $V = \frac{0.00454\,ft^3/s}{0.0216\,ft^2}$

    $V = 0.210\,FPS$

    $Re = \frac{D \cdot V \cdot \rho}{\mu}$

    $Re = \frac{0.166\,ft \cdot 0.210\,ft/s \cdot 62.4\,lb_m/ft^3}{6.76 \cdot 10^{-4}\,lb_m/ft \cdot s}$

    $Re = 3{,}217.8$

    The correct answer is B.

19. $L_A = L_{A,\,pipe} + L_{A,\,coil} + L_{A,\,balance\ valve}$

$L_A = 30\ ft + 40\ ft + L_{A,\,balance\ valve}$

$L_A = 70\ ft + L_{A,\,balance\ valve}$

$L_B = L_{B,\,pipe} + L_{B,\,coil}$

$L_B = 160\ ft + 20\ ft$

$L_B = 180\ ft$

$$\frac{f \cdot L_A \cdot V_A^2}{2 \cdot D \cdot g} = \frac{f \cdot L_B \cdot V_B^2}{2 \cdot D \cdot g}$$

$L_A \cdot V_A^2 = L_B \cdot V_B^2$

Since $Q = V \cdot A$ and pipe diameter remains constant, $V \propto Q$

$L_A \cdot Q_A^2 = L_B \cdot Q_B^2$

$(70\ ft + L_{A,\,balance\ valve}) \cdot (30\ GPM)^2 = 180\ ft \cdot (20\ GPM)^2$

$L_{A,\,balance\ valve} = \dfrac{180\ ft \cdot (20\ GPM)^2}{(30\ GPM)^2} - 70\ ft$

$L_{A,\,balance\ valve} = 10\ ft \cdot \dfrac{0.433\ psi}{ft}$

$L_{A,\,balance\ valve} = 4.33\ psi$

The correct answer is A.

20. From pressure-enthalpy diagram for R-22,

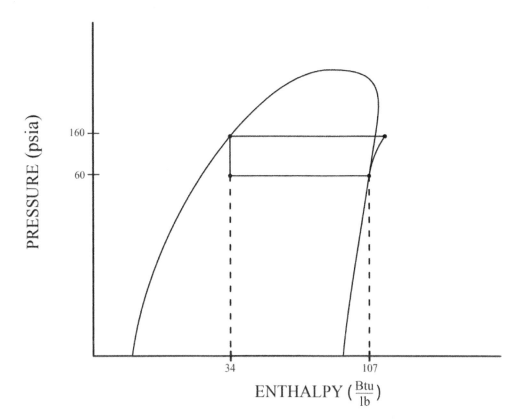

$h_1 = 107 \frac{\text{Btu}}{\text{lb}_m}$

$h_4 = 34 \frac{\text{Btu}}{\text{lb}_m}$

$\Delta h_{\text{evap}} = h_1 - h_4$

$\Delta h_{\text{evap}} = 107 \frac{\text{Btu}}{\text{lb}_m} - 34 \frac{\text{Btu}}{\text{lb}_m}$

$\Delta h_{\text{evap}} = 73 \frac{\text{Btu}}{\text{lb}_m}$

$Q_{\text{in}} = 2.5 \text{ Tons} \cdot \frac{12{,}000 \text{ Btu}/\text{hr}}{\text{Ton}}$

$Q_{\text{in}} = 30{,}000 \frac{\text{Btu}}{\text{hr}}$

$\dot{m} = \frac{Q_{\text{in}}}{\Delta h_{\text{evap}}}$

$$\dot{m} = \frac{30{,}000 \; \text{Btu}/\text{hr}}{73 \; \text{Btu}/\text{lb}_m}$$

$$\dot{m} = 411 \; \frac{\text{lb}_m}{\text{hr}}$$

The correct answer is D.

21. From ASHRAE Psychrometric Chart #1, the correct answer is A.

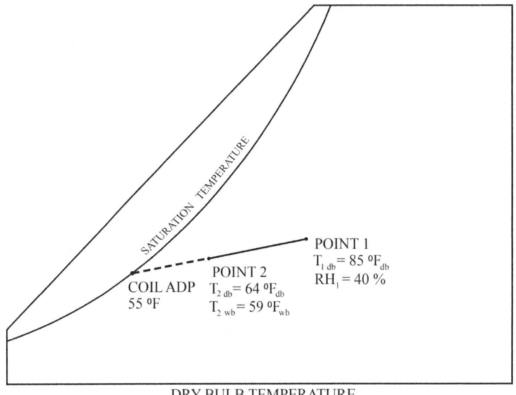

22. $\text{COP}_{HP} = \dfrac{T_H}{T_H - T_C}$

$\text{COP}_{HP} = \dfrac{(103\,°F + 460\,°R)}{103\,°F - 47\,°F}$

$\text{COP}_{HP} = 10.05$

The correct answer is C.

23. $A = \pi \cdot (\frac{6 \text{ in}}{2 \cdot 12 \text{ in}/\text{ft}})^2$

   $A = 0.196 \text{ ft}^2$

   $Q = V \cdot A$

   $Q = 15 \frac{\text{ft}}{\text{s}} \cdot 0.196 \text{ ft}^2$

   $Q = 2.94 \frac{\text{ft}^3}{\text{s}} \cdot 448.8 \frac{\text{GPM}}{\text{ft}^3/\text{s}}$

   $Q = 1,319.5 \text{ GPM}$

   $P_{hp} = \frac{h \cdot Q \cdot SG}{3956 \cdot \eta}$

   $P_{hp} = \frac{30 \text{ ft} \cdot 1,319.5 \text{ GPM} \cdot 1.0}{3956 \cdot 0.67 \cdot 0.90}$

   $P_{hp} = 16.6 \text{ HP}$

   $P_{kW} = 16.6 \text{ HP} \cdot 0.745 \frac{\text{kW}}{\text{HP}}$

   $P_{kW} = 12.4 \text{ kW}$

   The correct answer is C.

24. $A_{14'' \text{ square}} = 14 \text{ in} \cdot 14 \text{ in} \cdot \frac{1 \text{ ft}^2}{144 \text{ in}^2}$

   $A_{14'' \text{ square}} = 1.36 \text{ ft}^2$

   $A_{12'' \text{ round}} = \pi \cdot (\frac{12 \text{ in}}{2 \cdot 12 \text{ in}/\text{ft}})^2$

   $A_{12'' \text{ round}} = 0.785 \text{ ft}^2$

   $A_{10'' \text{ round}} = \pi \cdot (\frac{10 \text{ in}}{2 \cdot 12 \text{ in}/\text{ft}})^2$

   $A_{10'' \text{ round}} = 0.545 \text{ ft}^2$

   $Q = V \cdot A$

$$Q_1 = 1.36 \text{ ft}^2 \cdot 800 \frac{\text{ft}}{\text{min}}$$

$$Q_1 = 1{,}088 \frac{\text{ft}^3}{\text{min}}$$

$$Q_2 = 0.785 \text{ ft}^2 \cdot 500 \frac{\text{ft}}{\text{min}}$$

$$Q_2 = 392.5 \frac{\text{ft}^3}{\text{min}}$$

$$Q_3 = 0.545 \text{ ft}^2 \cdot 500 \frac{\text{ft}}{\text{min}}$$

$$Q_3 = 272.5 \frac{\text{ft}^3}{\text{min}}$$

$$Q_4 = 0.545 \text{ ft}^2 \cdot 550 \frac{\text{ft}}{\text{min}}$$

$$Q_4 = 300 \frac{\text{ft}^3}{\text{min}}$$

$$Q_5 = Q_1 - Q_2 - Q_3 - Q_4$$

$$Q_5 = 1{,}088 \frac{\text{ft}^3}{\text{min}} - 392.5 \frac{\text{ft}^3}{\text{min}} - 272.5 \frac{\text{ft}^3}{\text{min}} - 300 \frac{\text{ft}^3}{\text{min}}$$

$$Q_5 = 123 \frac{\text{ft}^3}{\text{min}}$$

$$V_5 = \frac{Q}{A}$$

$$V_5 = \frac{123 \text{ ft}^3/\text{min}}{0.545 \text{ ft}^2}$$

$$V_5 = 226 \text{ FPM}$$

The correct answer is C.

25. $Q = 10 \text{ kW} \cdot 3{,}412.1 \frac{\text{Btu}/\text{hr}}{\text{kW}}$

$$Q = 34{,}121 \frac{\text{Btu}}{\text{hr}}$$

$$Q = 1.08 \cdot \text{CFM} \cdot \Delta T$$

$$T_{2,DB} = \frac{Q}{1.08 \cdot CFM} + T_1$$

$$T_{2,DB} = \frac{34{,}121 \, \text{Btu/hr}}{1.08 \cdot 2{,}560 \, \text{CFM}} + 55°F$$

$$T_{2,DB} = 67.3 \, °F$$

From ASHRAE Psychrometric Chart #1,

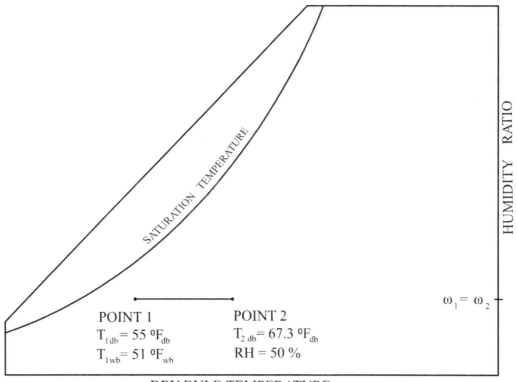

$RH_2 = 50\%$

The correct answer is D.

26. $Q = 1.08 \cdot CFM \cdot \Delta T$

$$T_2 = \frac{Q}{1.08 \cdot CFM} + T_1$$

$$T_2 = \frac{44 \, \text{kW} \cdot 3{,}412.1 \, \frac{\text{Btu/hr}}{\text{kW}}}{1.08 \cdot 3{,}000} + 60 \, °F$$

$$T_2 = 106.3 \, °F$$

$$\dot{m}_{air} = \frac{V}{\nu}$$

$$\dot{m}_{air} = \frac{3{,}000 \text{ ft}^3/\text{min}}{13.15 \text{ ft}^3/\text{lb}} \cdot 60 \frac{\text{min}}{\text{hr}}$$

$$\dot{m}_{air} = 13{,}688 \frac{\text{lb}}{\text{hr}}$$

$$\dot{m}_{water} = 22 \frac{\text{gal}}{\text{hr}} \cdot 0.1337 \frac{\text{ft}^3}{\text{gal}} \cdot 62.4 \frac{\text{lb}}{\text{ft}^3}$$

$$\dot{m}_{water} = 183.5 \frac{\text{lb}}{\text{hr}}$$

From ASHRAE Psychrometric Chart #1, $\omega_2 = 0.0022 \frac{\text{lb}_w}{\text{lb}_a}$

$$\dot{m}_w = \dot{m}_a \cdot (\omega_3 - \omega_2)$$

$$\omega_3 = \frac{\dot{m}_w}{\dot{m}_a} + \omega_2$$

$$\omega_3 = \frac{183.5 \text{ lb}/\text{hr}}{13{,}500 \text{ lb}/\text{hr}} + 0.0022 \frac{\text{lb}_w}{\text{lb}_a}$$

$$\omega_3 = 0.0158 \frac{\text{lb}_w}{\text{lb}_a}$$

From ASHRAE Psychrometric Chart #1, $RH_3 = 67\%$

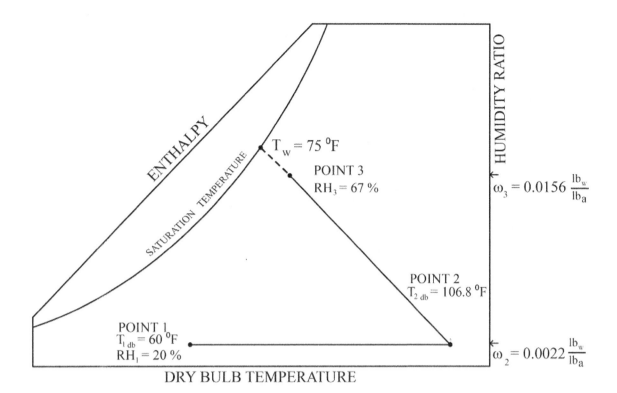

The correct answer is A.

27. $\dot{m}_{air} = \dfrac{\dot{V}}{v}$

$\dot{m}_{air} = \dfrac{1{,}400 \,\text{ft}^3/\text{min}}{13.85 \,\text{ft}^3/\text{lb}} \cdot 60 \,\dfrac{\text{min}}{\text{hr}}$

$\dot{m}_{air} = 6{,}065 \,\dfrac{\text{lb}}{\text{hr}}$

From ASHRAE Psychrometric Chart #1,

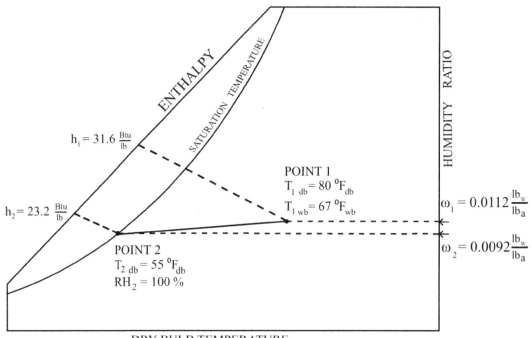

$h_1 = 31.6 \dfrac{\text{Btu}}{\text{lb}_m}$

$h_2 = 23.2 \dfrac{\text{Btu}}{\text{lb}_m}$

$\omega_1 = 0.0112 \dfrac{\text{lb}_w}{\text{lb}_a}$

$\omega_2 = 0.0092 \dfrac{\text{lb}_w}{\text{lb}_a}$

From steam tables,

$h_f = 28.08 \dfrac{\text{Btu}}{\text{lb}_m}$

$Q = \dot{m}_a \cdot [(h_1 - h_2) - h_f \cdot (\omega_1 - \omega_2)]$

$Q = 6{,}065 \dfrac{\text{lbm}}{\text{hr}} \cdot [(31.6 \dfrac{\text{Btu}}{\text{lb}_m} - 23.2 \dfrac{\text{Btu}}{\text{lb}_m}) - 28.08 \dfrac{\text{Btu}}{\text{lb}_m} \cdot (0.0112 \dfrac{\text{lb}_w}{\text{lb}_a} - 0.0092 \dfrac{\text{lb}_w}{\text{lb}_a})]$

$Q = 50{,}642 \dfrac{\text{Btu}}{\text{hr}}$

The correct answer is B.

28. $SHR = \dfrac{Q_S}{Q_T}$

$SHR = \dfrac{30{,}000 \, \text{Btu}/\text{hr}}{42{,}600 \, \text{Btu}/\text{hr}}$

$SHR = 0.70$

$BF = \dfrac{T_{db,out} - T_{coil}}{T_{db,in} - T_{coil}}$

$T_{db,out} = BF \cdot (T_{db,in} - T_{coil}) + T_{coil}$

$T_{db,out} = 0.2 \cdot (75\,°F - 52\,°F) + 52\,°F$

$T_{db,out} = 56.6\,°F$

From ASHRAE Psychrometric Chart #1, $\omega_1 = 0.0074 \, \dfrac{lb_w}{lb_a}$

SHR intersects $T_{db,out}$ at $\omega_2 = 0.0054 \, \dfrac{lb_w}{lb_a}$

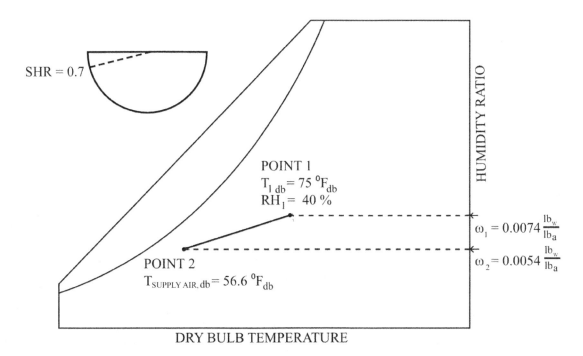

$\dot{m}_{cond} = \dfrac{V_{air}}{v_{air}} \cdot (\omega_1 - \omega_2)$

$\dot{m}_{cond} = \dfrac{200 \, \text{ft}^3/\text{min}}{13.65 \, \text{ft}^3/\text{lb}} \cdot \left(0.0074 \, \dfrac{lb_w}{lb_a} - 0.0054 \, \dfrac{lb_w}{lb_a}\right)$

$$\dot{m}_{cond} = 0.029 \frac{lb_w}{min} \cdot \frac{1 \, ft^3}{62.4 \, lb_w} \cdot \frac{7.48 \, gal}{1 \, ft^3} \cdot \frac{60 \, min}{hr}$$

$$\dot{m}_{cond} = 0.211 \frac{gal}{hr}$$

The correct answer is A.

29. $x = \dfrac{\dot{m}_{vap}}{\dot{m}_{vap} + \dot{m}_{liq}}$

$x_{return} = \dfrac{80 \, lb_m}{100 \, lb_m}$

$x_{return} = 0.8$

$p_{absolute} = p_{guage} + 14.7 \, psia$

$p_{absolute} = 24.7 \, psia$

From saturated steam tables at p = 25 psia

$s_f = 0.3535 \dfrac{Btu}{lb_m \cdot °R}$

$s_{fg} = 1.3606 \dfrac{Btu}{lb_m \cdot °R}$

$s_{mix} = s_f + x \cdot s_{fg}$

$s_{mix} = 0.3535 \dfrac{Btu}{lb_m \cdot °R} + 0.8 \cdot 1.3606 \dfrac{Btu}{lb_m \cdot °R}$

$s_{mix} = 1.44 \dfrac{Btu}{lb_m \cdot °R}$

The correct answer is C.

30. From ASHRAE Psychrometric Chart #1,

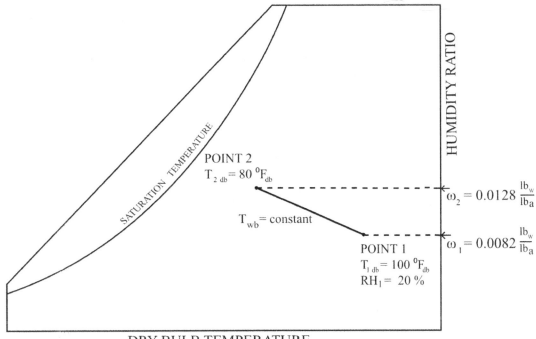

$\omega_1 = 0.0128 \frac{lb_w}{lb_a}$

$\omega_2 = 0.0082 \frac{lb_w}{lb_a}$

$\dot{m}_w = \frac{V_{air}}{v_{air}} \cdot (\omega_1 - \omega_2)$

$\dot{m}_w = \frac{5{,}000 \, ft^3/min}{14.3 \, ft^3/lb} \cdot (0.0128 \frac{lb_w}{lb_a} - 0.0082 \frac{lb_w}{lb_a})$

$\dot{m}_w = 1.6 \frac{lb}{min} \cdot \frac{1 \, ft^3}{62.4 \, lb_w} \cdot \frac{7.48 \, gal}{1 \, ft^3} \cdot \frac{60 \, min}{hr}$

$\dot{m}_w = 11.5 \, GPH$

The correct answer is B.

31. The First Law of Thermodynamics states that energy cannot be created or destroyed. The correct answer is C.

32. From ASHRAE Psychrometric Chart #1, draw a line between the room air temperature and supply air temperature, then transcribe up to the SHR scale.

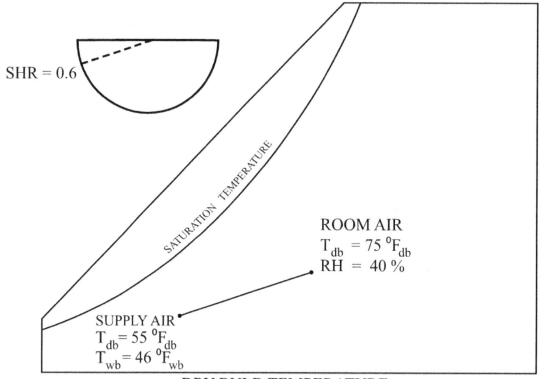

The correct answer is B.

33. $Q_{GPM} = C_v \cdot \sqrt{\dfrac{\Delta p_{psi}}{SG}}$

$\Delta p_{psi} = \dfrac{Q_{GPM}^2}{C_v^2} \cdot SG$

$\Delta p_{psi} = \dfrac{60 \text{ GPM}^2}{16^2} \cdot 1$

$\Delta p_{psi} = 14.06 \text{ psi}$

$p_2 = 55 \text{ psi} - 14.06 \text{ psi}$

$p_2 = 40.9 \text{ psi}$

The correct answer is B.

34. $\frac{Q}{L} = \frac{2 \cdot \pi \cdot k_1}{\ln(R_2/R_1)} \cdot (T_1 - T_2) + \frac{2 \cdot \pi \cdot k_2}{\ln(R_3/R_2)} \cdot (T_2 - T_3)$

$k_2 = \frac{\Delta x}{R}$

$k_2 = \frac{2 \text{ in} \cdot \frac{1 \text{ ft}}{12 \text{ in}}}{5.0 \frac{°F \cdot ft^2 \cdot hr}{Btu}}$

$k_2 = 0.033 \frac{Btu}{hr \cdot °F \cdot ft}$

$R_1 = \frac{2.9 \text{ in} \cdot \frac{1 \text{ ft}}{12 \text{ in}}}{2}$

$R_1 = 0.121 \text{ ft}$

$R_2 = \frac{3.5 \text{ in} \cdot \frac{1 \text{ ft}}{12 \text{ in}}}{2}$

$R_2 = 0.146 \text{ ft}$

$R_3 = R_2 + t$

$R_3 = 0.146 \text{ ft} + \left(2 \text{ in} \cdot \frac{1 \text{ ft}}{12 \text{ in}}\right)$

$R_3 = 0.3125 \text{ ft}$

$\frac{Q}{L} = \frac{2 \cdot \pi \cdot 20 \frac{Btu}{hr \cdot °F \cdot ft}}{\ln(0.146 \text{ ft}/0.121 \text{ ft})} \cdot (160\,°F - 158\,°F) + \frac{2 \cdot \pi \cdot 0.033 \frac{Btu}{hr \cdot °F \cdot ft}}{\ln(0.3125 \text{ ft}/0.146 \text{ ft})} \cdot (158\,°F - 82\,°F)$

$\frac{Q}{L} = 1{,}358 \frac{Btu}{hr \cdot ft}$

The correct answer is B.

35. $\frac{f_1 \cdot L_1 \cdot V_1^2}{2 \cdot D_1 \cdot g} = \frac{f_2 \cdot L_2 \cdot V_2^2}{2 \cdot D_2 \cdot g}$

$\frac{0.001 \text{ in}/\text{in} \cdot 18 \text{ ft} \cdot V_1^2}{2 \cdot 5 \text{ in} \cdot 32.2 \text{ ft}/s^2} = \frac{0.0008 \text{ in}/\text{in} \cdot 36 \text{ ft} \cdot V_2^2}{2 \cdot 3 \text{ in} \cdot 32.2 \text{ ft}/s^2}$

$V_1 = 1.63 \cdot V_2$

$$A_1 = \pi \cdot \left(\frac{5 \text{ in} \cdot {}^{1 \text{ ft}}/_{12 \text{ in}}}{2}\right)^2$$

$$A_1 = 0.136 \text{ ft}^2$$

$$A_2 = \pi \cdot \left(\frac{3 \text{ in} \cdot {}^{1 \text{ ft}}/_{12 \text{ in}}}{2}\right)^2$$

$$A_2 = 0.049 \text{ ft}^2$$

$$Q_1 + Q_2 = Q_T$$

$$A_1 \cdot V_1 + A_2 \cdot V_2 = A_T \cdot V_T$$

$$0.136 \text{ ft}^2 \cdot V_1 + 0.049 \text{ ft}^2 \cdot V_2 = 0.136 \text{ ft}^2 \cdot 15 \, \frac{\text{ft}}{\text{min}}$$

$$0.136 \text{ ft}^2 \cdot V_1 + 0.049 \text{ ft}^2 \cdot V_2 = 2.04 \, \frac{\text{ft}^3}{\text{min}}$$

Substituting $V_1 = 1.63 \cdot V_2$ gives:

$$0.136 \text{ ft}^2 \cdot (1.63 \cdot V_2) + 0.049 \text{ ft}^2 \cdot V_2 = 2.04 \, \frac{\text{ft}^3}{\text{min}}$$

$$V_2 = 7.53 \, \frac{\text{ft}}{\text{min}}$$

$$Q_2 = V_2 \cdot A_2$$

$$Q_2 = 7.53 \, \frac{\text{ft}}{\text{min}} \cdot 0.049 \text{ ft}^2 \cdot 7.48 \, \frac{\text{gal}}{\text{ft}^3}$$

$$Q_2 = 2.77 \, \frac{\text{ft}^3}{\text{min}}$$

The correct answer is B.

36. $W_{in} = 5 \text{ kW} \cdot 3{,}412.1 \, \frac{{}^{\text{Btu}}/_{\text{hr}}}{\text{kW}}$

$$W_{in} = 17{,}060 \, \frac{\text{Btu}}{\text{hr}}$$

$$COP = \frac{Q_{in}}{W_{in}}$$

$$COP = \frac{27{,}000 \, {}^{\text{Btu}}/_{\text{hr}}}{17{,}060 \, {}^{\text{Btu}}/_{\text{hr}}}$$

COP = 1.58

The correct answer is A.

37. The correct answer is B.

38. $p_a = p_g + 14.7$ psi

$p_a = 5$ psig $+ 14.7$ psi

$p_a = 19.7$ psia

Round to 20 psia

From superheated steam tables at p = 20 psia, t = 400 °F,

$h_g = 1{,}239.3 \frac{\text{Btu}}{\text{lb}_m}$

$v = 25.43 \frac{\text{ft}^3}{\text{lb}_m}$

From saturated steam tables at p = 20 psia,

$h_f = 196.3 \frac{\text{Btu}}{\text{lb}_m}$

$\dot{m} = \frac{Q}{h_g - h_f}$

$\dot{m} = \frac{116{,}000 \text{ Btu}/\text{hr}}{1{,}239.3 \text{ Btu}/\text{lb}_m - 196.3 \text{ Btu}/\text{lb}_m}$

$\dot{m} = 111.2 \frac{\text{lb}_m}{\text{hr}}$

$r = 0.0625$ ft ← .073

$A = \pi \cdot (0.0625 \text{ ft})^2$

$A = 0.0123$ ft² → 0.0176²

$\dot{m} = \rho \cdot V \cdot A$

$\dot{m} = \frac{V \cdot A}{v}$

$$V = \frac{\dot{m} \cdot v}{A}$$

$$V = \frac{111.2 \, \text{lb}_m/\text{hr} \cdot 25.43 \, \text{ft}^3/\text{lb}_m}{0.0123 \, \text{ft}^2} \cdot \frac{1 \, \text{hr}}{3{,}600 \, \text{s}}$$

$V = 63.9$ FPS → 46...

The correct answer is A.

39. $r = 0.125$ ft

$A = \pi \cdot (0.125 \, \text{ft})^2$

$A = 0.049$ ft²

$$Q = 90 \, \frac{\text{gal}}{\text{min}} \cdot \frac{1 \, \text{min}}{60 \, \text{s}} \cdot \frac{1 \, \text{ft}^3}{7.48 \, \text{gal}}$$

$Q = 0.20 \, \frac{\text{ft}^3}{\text{s}}$

$V = \frac{Q}{A}$

$V = \frac{0.2 \, \text{ft}^3/\text{s}}{0.049 \, \text{ft}^2}$

$V = 4.08 \, \frac{\text{ft}}{\text{s}}$

$p_{\text{Total}} = p_{\text{water hammer}} + p_{\text{height}} + p_{\text{guage}}$

$p_{\text{w.h.}} = \frac{\rho \cdot c_s \cdot V}{g_c}$

$$p_{\text{w.h.}} = \frac{62.4 \, \text{lb}/\text{ft}^3 \cdot 4{,}720 \, \text{ft}/\text{s} \cdot 4.08 \, \text{ft}/\text{s}}{32.2 \, \text{ft}/\text{s}^2}$$

$p_{\text{w.h.}} = 37{,}319 \, \frac{\text{lb}}{\text{ft}^2} \cdot \frac{1 \, \text{ft}^2}{144 \, \text{in}^2}$

$p_{\text{w.h.}} = 259$ psi

$p_z = 400 \, \text{ft} \cdot 0.433 \, \frac{\text{psi}}{\text{ft}}$

$p_z = 173$ psi

p_T = 259 psi + 173 psi + 100 psi

p_T = 532 psi

The correct answer is D.

40. From pressure-enthalpy diagram for R-134a, x = 0.55.

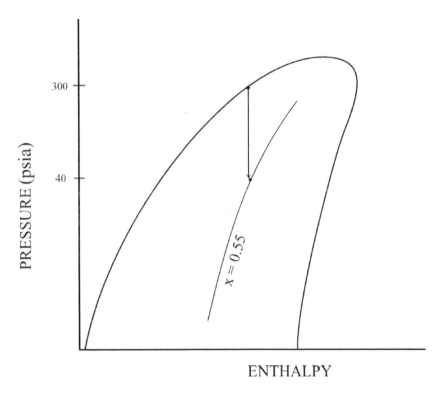

The correct answer is A.

41. $COP = \dfrac{Q_{in}}{W_{in}}$

$COP = \dfrac{h_1 - h_4}{h_2 - h_1}$

$COP = \dfrac{133\ Btu/lb - 75\ Btu/lb}{147\ Btu/lb - 133\ Btu/lb}$

$COP = 4.14$

The correct answer is C.

42. Solution 1

From ASHRAE Psychrometric Chart #1, RH = 50%

From steam tables at T = 72 °F, p_{sat} = 0.3889 psia

$RH = \dfrac{p_{partial}}{p_{sat}}$

$p_{partial} = 0.50 \cdot 0.3889$ psia

$p_{partial} = 0.194$ psia

The correct answer is A.

Alternate Solution

From ASHRAE Psychrometric Chart #1, T_{dp} = 52 °F

From steam tables at T = 52 °F_{db}, p_{sat} = 0.1918 psia

43. Plot 90 °F_{db} / 42% RH on ASHRAE Psychrometric Chart #1, read T_{dp} = 64 °F on a horizontal line left of the point.

The correct answer is B.

44. Standard air conditions are T = 70 °F and p = 14.7 psia

$K_d = \dfrac{p_{std}}{p_{act}} \cdot \dfrac{T_{act}}{T_{std}}$

$K_d = \dfrac{14.7 \text{ psia}}{11.8 \text{ psia}} \cdot \dfrac{85 \text{ °F} + 460 \text{ R}}{70 \text{ °F} + 460 \text{ R}}$

$K_d = 1.28$

$\text{ACFM} = K_d \cdot \text{SCFM}$

$\text{ACFM} = 1.28 \cdot 2{,}500 \text{ CFM}$

$\text{ACFM} = 3{,}202 \text{ CFM}$

The correct answer is C.

45. From pressure enthalpy diagram for R-22,

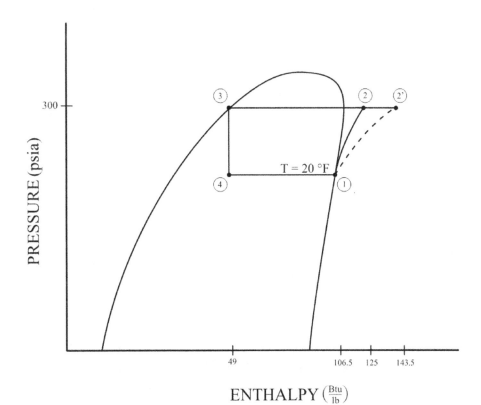

$$h_4 = 49 \frac{Btu}{lb}$$

$$h_1 = 106.5 \frac{Btu}{lb}$$

$$h_2 = 125 \frac{Btu}{lb}$$

$$h'_2 = h_1 + \frac{h_2 - h_1}{\eta}$$

$$h'_2 = 106.5 \frac{Btu}{lb} + \frac{125 \, Btu/lb - 106.5 \, Btu/lb}{0.5}$$

$$h'_2 = 143.5 \frac{Btu}{lb}$$

$$Q_{in} = \dot{m} \cdot (h_1 - h_4)$$

$$\dot{m} = \frac{70{,}000 \, Btu/hr}{106.5 \, Btu/lb - 49 \, Btu/lb}$$

$$\dot{m} = 1{,}217 \frac{lb}{hr}$$

$$W = \dot{m} \cdot (h'_2 - h_1)$$

$$W = 1{,}217 \frac{lb}{hr} \cdot \left(143.5 \frac{Btu}{lb} - 106.5 \frac{Btu}{lb}\right)$$

$$W = 45{,}029 \frac{Btu}{hr} \cdot \frac{1 \, kW}{3{,}412.1 \, Btu/hr}$$

$$W = 13.2 \, kW$$

The correct answer is B.

46. $T_{mix,db} = T_{a,db} + \left(\frac{\dot{v}_b}{\dot{v}_a + \dot{v}_b}\right) \cdot (T_{b,db} - T_{a,db})$

$T_{mix,db} = 60 \,°F_{db} + \left(\frac{15{,}000 \, CFM}{8{,}000 \, CFM + 15{,}000 \, CFM}\right) \cdot (73 \,°F_{db} - 60 \,°F_{db})$

$T_{mix,db} = 68.5 \,°F_{db}$

From ASHRAE Psychrometric Chart #1, $T_{mix,wb} = 56 \,°F_{,wb}$.

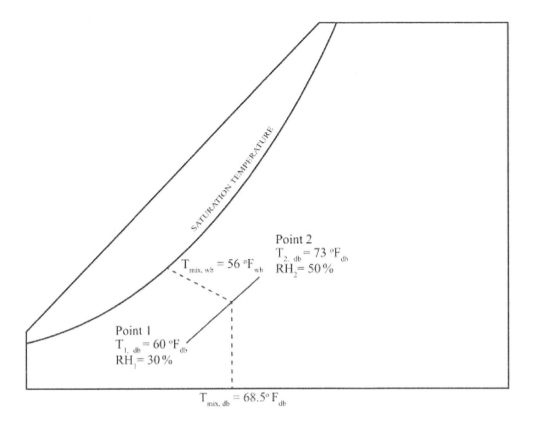

The correct answer is A.

47. $\dot{m}_{hot\ water} = 25\ \frac{gal}{min} \cdot \frac{ft^3}{7.48\ gal} \cdot 62.4\ \frac{lb}{ft^3} \cdot 60\ \frac{min}{hr}$

$\dot{m}_{hot\ water} = 12{,}513\ \frac{lb}{hr}$

$\dot{m}_{cold\ water} = 60\ \frac{gal}{min} \cdot \frac{ft^3}{7.48\ gal} \cdot 62.4\ \frac{lb}{ft^3} \cdot 60\ \frac{min}{hr}$

$\dot{m}_{cold\ water} = 30{,}032\ \frac{lb}{hr}$

$C_r = \frac{C_{min}}{C_{max}}$

$C_r = \frac{\dot{m}_{hot\ water} \cdot C_{hot}}{\dot{m}_{cold\ water} \cdot C_{cold}}$

$C_r = \frac{12{,}513\ lb/hr \cdot 1\ Btu/lb \cdot °F}{30{,}032\ lb/hr \cdot 1\ Btu/lb \cdot °F}$

$C_r = 0.416$

85

$$\text{NTU} = \frac{U \cdot A}{C_{min}}$$

$$\text{NTU} = \frac{44 \, \text{Btu}/\text{hr} \cdot \text{ft}^2 \cdot °F \cdot 148 \, \text{ft}^2}{12{,}513 \, \text{Btu}/\text{hr} \cdot °F}$$

$$\text{NTU} = 0.52$$

For a counter flow heat exchanger,

$$\varepsilon = \frac{1 - \exp[-\text{NTU}(1 - C_r)]}{1 - C_r \cdot \exp[-\text{NTU}(1 - C_r)]}$$

$$\varepsilon = \frac{1 - \exp[-0.52(1 - 0.416)]}{1 - 0.416 \cdot \exp[-0.52(1 - 0.416)]}$$

$$\varepsilon = 0.378$$

$$Q_{max} = C_{min} \cdot (t_{hot,\,in} - t_{cold,\,in})$$

$$Q_{max} = 12{,}513 \, \frac{\text{Btu}}{\text{hr} \cdot °F} \cdot (300 \, °F - 80 \, °F)$$

$$Q_{max} = 2{,}752{,}860 \, \frac{\text{Btu}}{\text{hr}}$$

$$Q = \varepsilon \cdot Q_{max}$$

$$Q = 0.378 \cdot 2{,}752{,}860 \, \frac{\text{Btu}}{\text{hr}}$$

$$Q = 1{,}040{,}580 \, \frac{\text{Btu}}{\text{hr}}$$

$$t_{cold,\,out} = t_{cold,\,in} + \frac{Q}{\dot{m} \cdot c_p}$$

$$t_{cold,\,out} = 80 \, °F + \frac{1{,}040{,}580 \, \text{Btu}/\text{hr}}{30{,}032 \, \text{lb}/\text{hr} \cdot 1 \, \text{Btu}/\text{lb} \cdot °F}$$

$$t_{cold,\,out} = 114.6 \, °F$$

The correct answer is A.

48. From saturated steam tables at T = 280°F,

$$s_f = 0.4099 \, \frac{\text{Btu}}{\text{lb}_m \cdot °R}$$

$s_g = 1.6601 \frac{Btu}{lb_m \cdot °R}$

$h_f = 249.2 \frac{Btu}{lb}$

$h_{fg} = 924.7 \frac{Btu}{lb}$

$x = \frac{s_{mix} - s_f}{s_g - s_f}$

$x = \frac{0.762 \, Btu/lb_m \cdot °R - 0.4099 \, Btu/lb_m \cdot °R}{1.6601 \, Btu/lb_m \cdot °R - 0.4099 \, Btu/lb_m \cdot °R}$

$x = 0.282$

$h_{mix} = h_f + x \cdot h_{fg}$

$h_{mix} = 249.2 \frac{Btu}{lb} + 0.282 \cdot 924.7 \frac{Btu}{lb}$

$h_{mix} = 510 \frac{Btu}{lb}$

The correct answer is A.

49. $W = 400 \text{ kW} \cdot 3{,}412.1 \frac{Btu/hr}{kW}$

$W = 1{,}364{,}840 \frac{Btu}{hr}$ → 1365200

From superheated steam tables at p = 5 psia and T = 300°F, $h_i = 1{,}194.8 \frac{Btu}{lb}$

$W = \dot{m} \cdot (h_i - h_e)$

$h_e = h_i - \frac{W}{\dot{m}}$

$h_e = 1{,}194.8 \frac{Btu}{lb} - \frac{1{,}364{,}840 \, Btu/hr}{15 \, lb/s \cdot 60 \, s/min \cdot 60 \, min/hr}$

$h_e = 1{,}169.5 \frac{Btu}{lb}$

From superheated steam tables at p = 1 psia and T = 300 °F, h = 1,195.7 $\frac{Btu}{lb}$ and at T = 200 °F, h = 1,150.1 $\frac{Btu}{lb}$. Interpolate to find t_e.

$$t_e = 200 \text{ °F} + \left(1{,}169.5 \frac{Btu}{lb} - 1{,}150.1 \frac{Btu}{lb}\right)\left(\frac{300 \text{ °F} - 200 \text{ °F}}{1{,}195.7 \text{ }Btu/lb - 1{,}150.1 \text{ }Btu/lb}\right)$$

$t_e = 242.5$ °F

The correct answer is A.

Answer = B

50. $T_1 = 80$ °F $+ 460 = 540$ R

$p_1 = 5$ psig $+ 14.7$ psi $= 19.7$ psia

$p_2 = 20$ psig $+ 14.7$ psi $= 34.7$ psia

For an ideal gas undergoing isentropic compression/ expansion,

$$T_1 \cdot p_1^{1-k/k} = T_2 \cdot p_2^{1-k/k}$$

$$T_2 = \frac{T_1 \cdot p_1^{1-k/k}}{p_2^{1-k/k}}$$

$$T_2 = \frac{540 \text{ R} \cdot (19.7 \text{ psia})^{1-1.4/1.4}}{(34.7 \text{ psia})^{1-1.4/1.4}}$$

$T_2 = 634.8$ °R

$T_2 = 174.8$ °F

The correct answer is A.

51. <u>Solution 1</u>

$Q_L = Q_T - Q_S$

$Q_L = (4.5 \cdot CFM \cdot \Delta h) - (1.08 \cdot CFM \cdot \Delta T)$

From ASHRAE Psychrometric Chart #1 and #2, $h_1 = 9.8 \frac{Btu}{lb}$, $h_2 = 26.2 \frac{Btu}{lb}$

$Q_L = (4.5 \cdot 2{,}000 \text{ CFM} \cdot 26.2 \frac{\text{Btu}}{\text{lb}} - 9.8 \frac{\text{Btu}}{\text{lb}}) - (1.08 \cdot 2{,}000 \text{ CFM} \cdot 73\,°F - 40\,°F)$

$Q_L = 76{,}320 \frac{\text{Btu}}{\text{lb}}$

Alternate Solution

$Q_L = 60 \cdot Q \cdot \rho \cdot \Delta\omega \cdot (1{,}061 + 0.444T)$

From ASHRAE Psychrometric Chart #1 and #2, $\omega_1 = 0.00051 \frac{\text{lb}_w}{\text{lb}_a}$, $\omega_2 = 0.0078 \frac{\text{lb}_w}{\text{lb}_a}$

$Q_L = 60 \cdot 2{,}000 \text{ CFM} \cdot \frac{1}{12.6 \text{ ft}^3/\text{lb}} \cdot (0.0078 \frac{\text{lb}_w}{\text{lb}_a} - 0.00051 \frac{\text{lb}_w}{\text{lb}_a}) \cdot (1{,}061 + 0.444(\frac{73\,°F + 40\,°F}{2}))$

$Q_L = 75{,}405 \frac{\text{Btu}}{\text{lb}}$

The correct answer is B.

52. $\omega = \frac{0.622 \cdot P_v}{P - P_v}$

$0.013 \frac{\text{lb}_w}{\text{lb}_a} = \frac{0.622 \cdot P_v}{12.5 \text{ psia} - P_v}$

$P_v = 0.256 \text{ psia}$

The correct answer is B.

SOLUTIONS PART II - APPLICATIONS

53. $Q = \dfrac{A \cdot \Delta T}{\Sigma R}$

   $8{,}000 \, \dfrac{\text{Btu}}{\text{hr}} = \dfrac{2{,}000 \text{ ft}^2 \cdot (72\,°\text{F} - 25\,°\text{F})}{(0.68 + 0.95 + R_{ins} + 0.5 + 0.68)\,°\text{F} \cdot \text{ft}^2 \cdot \text{hr}/\text{Btu}}$

   $R = 8.94 \, \dfrac{°\text{F} \cdot \text{ft}^2 \cdot \text{hr}}{\text{Btu}}$

   The correct answer is C.

54. From ASHRAE 62.1, Table 6-1

   $V_{bz} = R_p \cdot P_z + R_a \cdot A_z$

   For art room, $V_{art} = 30 \text{ people} \cdot 10 \, \dfrac{\text{CFM}}{\text{person}} + 1{,}200 \text{ ft}^2 \cdot 0.18 \, \dfrac{\text{CFM}}{\text{ft}^2}$

   $V_{art} = 516 \text{ CFM}$

   For computer lab, $V_{lab} = 60 \text{ people} \cdot 10 \, \dfrac{\text{CFM}}{\text{person}} + 2{,}000 \text{ ft}^2 \cdot 0.12 \, \dfrac{\text{CFM}}{\text{ft}^2}$ [handwritten: .18, wrong answer]

   $V_{lab} = 840 \text{ CFM}$

   $V_{total} = 840 \text{ CFM} + 516 \text{ CFM}$

   $V_{total} = 1{,}356 \text{ CFM}$

   The correct answer is C.

55. From ASHRAE 62.1, Table 6-4

   $V = A \cdot R_{EXH}$

   $V = 800 \text{ ft}^2 \cdot 0.60 \, \dfrac{\text{CFM}}{\text{ft}^2}$

   $V = 480 \text{ CFM}$

   The correct answer is B.

56. From ASHRAE Handbook – Refrigeration, Chapter 19,

   Freezing Point of Strawberries: 30.6 °F

   c_p Above Freezing: $0.96 \, \dfrac{\text{Btu}}{\text{lb} \cdot °\text{F}}$

   c_p Below Freezing: $0.44 \, \dfrac{\text{Btu}}{\text{lb} \cdot °\text{F}}$

Latent Heat of Fusion (LHF): $132 \frac{Btu}{lb}$

$$Q = m \cdot c_{p\ Above\ Freezing} \cdot (T_{Initial} - T_{Freezing}) + m \cdot LHF + m \cdot c_{p\ Below\ Freezing} \cdot (T_{Freezing} - T_{Final})$$

$$m = \frac{c_{p\ Above} \cdot (T_{Initial} - T_{Freezing}) + LHF + c_{p\ Below} \cdot (T_{Freezing} - T_{Final})}{Q}$$

$$m = \frac{0.96\ Btu/lb \cdot °F \cdot (74\ °F - 30.6\ °F) + 132\ Btu/lb + 0.44\ Btu/lb \cdot °F \cdot (30.6\ °F - 27\ °F)}{20\ Tons \cdot 12{,}000\ \frac{Btu/hr}{Ton} \cdot 24\ hr}$$

m = 32,857 lbs

The correct answer is A.

57. $Q = m \cdot c_{p\ Above\ Freezing} \cdot (T_{Initial} - T_{Freezing}) + m \cdot LHF + m \cdot c_{p\ Below\ Freezing} \cdot (T_{Freezing} - T_{Final})$

$Q = 300\ \frac{lb}{hr} \cdot 1.0\ \frac{Btu}{lb \cdot °F} \cdot (70\ °F - 32\ °F) + 300\ \frac{lb}{hr} \cdot 144\ \frac{Btu}{lb} + 300\ \frac{lb}{hr} \cdot 0.49\ \frac{Btu}{lb \cdot °F} \cdot (32\ °F - 14\ °F)$

$Q = 57{,}246\ \frac{Btu}{hr}$

The correct answer is C.

58. From ASHRAE Fundamentals, Chapter 26,

Outdoor air film, assume summer conditions, $R = 0.25\ \frac{°F \cdot ft^2 \cdot hr}{Btu}$

Indoor air film, no airflow, $R = 0.68\ \frac{°F \cdot ft^2 \cdot hr}{Btu}$

Hardboard Siding, $R = 0.15\ \frac{°F \cdot ft^2 \cdot hr}{Btu}$

Medium density particle board, $k = 0.94\ \frac{Btu \cdot in}{°F \cdot ft^2 \cdot hr}$

$R = \frac{\Delta}{k} = \frac{0.5\ in}{0.94\ Btu \cdot in /°F \cdot ft^2 \cdot hr} = 0.53\ \frac{°F \cdot ft^2 \cdot hr}{Btu}$

Douglas Fir wood, $k = 0.95$ to $1.01\ \frac{Btu \cdot in}{°F \cdot ft^2 \cdot hr}$

$R = \frac{\Delta}{k} = \frac{3.5\ in}{0.98\ Btu \cdot in /°F \cdot ft^2 \cdot hr} = 3.57\ \frac{°F \cdot ft^2 \cdot hr}{Btu}$

$R_{avg} = A_1 \cdot R_1 + A_2 \cdot R_2$

$R_{avg} = 0.16 \cdot 3.57\ \frac{°F \cdot ft^2 \cdot hr}{Btu} + (1.0 - 0.16) \cdot 30\ \frac{°F \cdot ft^2 \cdot hr}{Btu}$

$R_{avg} = 25.77\ \frac{°F \cdot ft^2 \cdot hr}{Btu}$

Plaster board, $k = 1.1 \frac{Btu \cdot in}{°F \cdot ft^2 \cdot hr}$

$R = \frac{\Delta}{k} = \frac{0.5 \text{ in}}{1.1 \, Btu \cdot in / °F \cdot ft^2 \cdot hr} = 0.45 \frac{°F \cdot ft^2 \cdot hr}{Btu}$

$R_{tot} = (0.25 + 0.15 + 0.53 + 25.77 + 0.45 + 0.68) \frac{°F \cdot ft^2 \cdot hr}{Btu}$

$R_{tot} = 27.83 \frac{°F \cdot ft^2 \cdot hr}{Btu}$

$\frac{Q}{A} = \frac{\Delta T}{R}$

$\frac{Q}{A} = \frac{(98 - 75) °F}{27.83 \, °F \cdot ft^2 \cdot hr / Btu}$

$\frac{Q}{A} = 0.83 \frac{Btu/hr}{ft^2}$

The correct answer is A.

59. From ASHRAE HVAC Systems and Equipment, Chapter 46,

Linear thermal expansion for Carbon Steel at 60 °F = $0.46 \frac{in}{100 \text{ ft}}$

Linear thermal expansion for Carbon Steel at 300 °F = $2.35 \frac{in}{100 \text{ ft}}$

$\Delta = 75 \text{ ft} \cdot (2.35 - 0.46 \frac{in}{100 \text{ ft}})$

$\Delta = 1.4175 \text{ in}$

The correct answer is C.

60. From ASHRAE Fundamentals, Chapter 7,

$Gain = \frac{(PV_2 - PV_1)}{(CS_2 - CS_1)}$

$Gain = \frac{(1,100 \text{ CFM} - 850 \text{ CFM})}{(65\% - 40\%)}$

$Gain = 10 \frac{CFM}{\%}$

The correct answer is C.

61. From ASHRAE Fundamentals, Chapter 7, the correct answer is B.

62. $\omega = \sqrt{\dfrac{k \cdot g_c}{m}}$

$\omega = \sqrt{\dfrac{4 \cdot 400 \frac{lb_f}{in} \cdot 12 \frac{in}{ft} \cdot 32.2 \frac{ft \cdot lb_m}{lb_f \cdot s^2}}{700 \, lb_m}}$

$\omega = 29.7 \, \dfrac{Rad}{s}$

$f_{nat} = \dfrac{\omega}{2 \cdot \pi}$

$f_{nat} = \dfrac{29.7 \, Rad/s}{2 \cdot \pi}$

$f_{nat} = 4.73 \, Hz$

The correct answer is D.

63. $\Delta T_A = 130 \, °F - 60 \, °F = 70 \, °F$

$\Delta T_B = 90 \, °F - 85 \, °F = 5 \, °F$

$\Delta T_{LM} = \dfrac{\Delta T_A - \Delta T_B}{\ln(\Delta T_A / \Delta T_B)}$

$\Delta T_{LM} = \dfrac{70 \, °F - 5 \, °F}{\ln(70 \, °F / 5 \, °F)}$

$\Delta T_{LM} = 24.6 \, °F$

$Q = U \cdot A \cdot \Delta T_{LM}$

$Q = 1{,}200 \dfrac{Btu}{hr \cdot ft^2 \cdot °F} \cdot 20 \, ft^2 \cdot 24.6 \, °F$

$Q = 590{,}400 \dfrac{Btu}{hr} \cdot \dfrac{1 \, Ton}{12{,}000 \, Btu/hr}$

$Q = 49.2 \, Tons$

The correct answer is B.

64. $\dot{m}_3 = 60 \, \frac{\text{gal}}{\text{hr}} \cdot \frac{1 \, \text{ft}^3}{7.48 \, \text{gal}} \cdot \frac{62.4 \, \text{lb}_m}{1 \, \text{ft}^3}$

$\dot{m}_3 = 500.5 \, \frac{\text{lb}_m}{\text{hr}}$

From saturated steam tables,

$h_1 = h_f + x \cdot h_{fg}$

$h_1 = 430.2 \, \frac{\text{Btu}}{\text{lb}} + 0.60 \cdot 774.9 \, \frac{\text{Btu}}{\text{lb}}$

$h_1 = 895.14 \, \frac{\text{Btu}}{\text{lb}}$

$h_2 = 148.04 \, \frac{\text{Btu}}{\text{lb}}$

$h_3 = 48.07 \, \frac{\text{Btu}}{\text{lb}}$

$Q_{in} = Q_{out}$

$\Sigma \, \dot{m}_{in} \cdot h_{in} = \Sigma \, \dot{m}_{out} \cdot h_{out}$

$\dot{m}_1 \cdot h_1 + \dot{m}_3 \cdot h_3 = \dot{m}_2 \cdot h_2 + \dot{m}_4 \cdot h_4$

$\dot{m}_1 = \dot{m}_2$

$\dot{m}_3 = \dot{m}_4$

$h_4 = \dfrac{\dot{m}_1 \cdot h_1 + \dot{m}_3 \cdot h_3 - \dot{m}_1 \cdot h_2}{\dot{m}_3}$

$h_4 = \dfrac{20 \, \frac{\text{lb}_m}{\text{hr}} \cdot 895.14 \, \frac{\text{Btu}}{\text{lb}} + 500.5 \, \frac{\text{lb}_m}{\text{hr}} \cdot 48.07 \, \frac{\text{Btu}}{\text{lb}} - 20 \, \frac{\text{lb}_m}{\text{hr}} \cdot 148.04 \, \frac{\text{Btu}}{\text{lb}}}{500.5 \, \frac{\text{lb}_m}{\text{hr}}}$

$h_4 = 77.9 \, \frac{\text{Btu}}{\text{lb}_m}$

From saturated steam tables,

$T_4 = 110°F$

The correct answer is A.

65. From ASHRAE Systems and Equipment, Chapter 46, 18" nominal schedule 40 steel pipe has an inside diameter of 16.876"

$$A = \pi \cdot \left(\frac{16.876 \text{ in}}{2 \cdot 12 \text{ in}/\text{ft}}\right)^2$$

$$A = 1.55 \text{ ft}^2$$

$$Q = 2,500 \frac{\text{gal}}{\text{min}} \cdot \frac{1 \text{ ft}^3}{7.48 \text{ gal}} \cdot \frac{1 \text{ min}}{60 \text{ s}}$$

$$Q = 5.57 \frac{\text{ft}^3}{\text{s}}$$

$$V = \frac{Q}{A}$$

$$V = \frac{5.57 \text{ ft}^3/\text{s}}{1.55 \text{ ft}^2}$$

$$V = 3.59 \frac{\text{ft}}{\text{s}}$$

$$h = \frac{f \cdot L \cdot V^2}{2 \cdot D \cdot g}$$

$$h = \frac{0.025 \cdot 300 \text{ ft} \cdot (3.59 \text{ ft}/\text{s})^2}{2 \cdot 1.406 \text{ ft} \cdot 32.2 \text{ ft}/\text{s}^2}$$

$$h = 1.067 \text{ ft}$$

The correct answer is C.

66. From ASHRAE Fundamentals, Chapter 20,

$$X = \frac{K_c \cdot Q_o}{V_x \cdot \sqrt{A_o}}$$

For high sidewall grilles, 0° deflection, $K_c = 5.7$

$Q_o = 2,000$ CFM

$V_x = 200$ FPM

$$A_o = 980 \text{ in}^2 \cdot \frac{1 \text{ ft}^2}{144 \text{ in}^2}$$

$$A_o = 6.8 \text{ ft}^2$$

$$X = \frac{5.7 \cdot 2{,}000 \text{ ft}^3/\text{min}}{200 \text{ ft}/\text{min} \cdot \sqrt{6.8 \text{ ft}^2}}$$

$$X = 21.8 \text{ ft}$$

The correct answer is B.

67. $$D_e = \frac{1.3 \cdot (\text{short side} \cdot \text{long side})^{5/8}}{(\text{short side} + \text{long side})^{1/4}}$$

$$D_e = \frac{1.3 \cdot (12 \text{ in} \cdot 30 \text{ in})^{5/8}}{(12 \text{ in} + 30 \text{ in})^{1/4}}$$

$$D_e = 20.2 \text{ in}$$

From standard friction loss in standard duct chart,

$$FL = 0.7 \frac{\text{in w.g.}}{100 \text{ ft}}$$

$$FL = 125 \text{ ft} \cdot 0.7 \frac{\text{in w.g.}}{100 \text{ ft}}$$

$$FL = 0.875 \text{ in w.g.}$$

The correct answer is C.

68. From ASHRAE Standard Designation and Safety Classification of Refrigerants, the correct answer is B.

69. $$A = 40 \text{ in} \cdot 30 \text{ in} \cdot \frac{1 \text{ ft}^2}{144 \text{ in}^2}$$

$$A = 8.33 \text{ ft}^2$$

$$V = \frac{Q}{A}$$

$$V = \frac{6,000 \text{ CFM}}{8.33 \text{ ft}^2}$$

$$V = 720 \text{ FPM}$$

$$FP = K \cdot \left(\frac{V_{FPM}}{4005}\right)^2$$

$$FP = 0.22 \cdot \left(\frac{720 \text{ FPM}}{4005}\right)^2$$

$$FP = 0.007 \text{ in w.g.}$$

The correct answer is A.

70. From ASHRAE HVAC Systems and Applications, Chapter 38, the correct answer is B.

71. $\varepsilon_s = \dfrac{\dot{m}_e \cdot c_{p,exhaust} \cdot (t_3 - t_4)}{C_{min} \cdot (t_3 - t_1)}$

$C_{min} = \dot{m}_{min} \cdot c_{p,min}$

From ASHRAE Psychrometric Chart #1,

$v_{outside\ air} = 12.6 \dfrac{ft^3}{lb}$

$v_{exhaust\ air} = 14.2 \dfrac{ft^3}{lb}$

$\dot{m}_{exhaust\ air} = 2,500 \dfrac{ft^3}{min} \cdot \dfrac{1 \text{ lb}}{14.2 \text{ ft}^3}$

$\dot{m}_{exhaust\ air} = 176 \dfrac{lb}{min}$

$\dot{m}_{min} = \dot{m}_{outside\ air} = 1,500 \dfrac{ft^3}{min} \cdot \dfrac{1 \text{ lb}}{12.6 \text{ ft}^3}$

$\dot{m}_{outside\ air} = 119 \dfrac{lb}{min}$

$$\varepsilon_s = \frac{176 \text{ lb}/\text{min} \cdot 0.24 \text{ Btu}/\text{lb} \cdot {}^\circ\text{F} \cdot (90 \, {}^\circ\text{F} - 77 \, {}^\circ\text{F})}{119 \text{ lb}/\text{min} \cdot 0.24 \text{ Btu}/\text{lb} \cdot {}^\circ\text{F} \cdot (90 \, {}^\circ\text{F} - 40 \, {}^\circ\text{F})}$$

$\varepsilon_s = 0.38$

The correct answer is B.

72. $Q_{\text{ammonia}} = \dot{m} \cdot c_{p,\text{ammonia}} \cdot \Delta T$

$\dot{m} = 100 \, \frac{\text{gal}}{\text{m}} \cdot 60 \, \frac{\text{min}}{\text{hr}} \cdot 0.13 \, \frac{\text{ft}^3}{\text{gal}} \cdot 35 \, \frac{\text{lb}}{\text{ft}^3}$

$\dot{m} = 27{,}300 \, \frac{\text{lb}}{\text{hr}}$

$Q_{\text{ammonia}} = 27{,}300 \, \frac{\text{lb}}{\text{hr}} \cdot 1.21 \, \frac{\text{Btu}}{\text{lb} \cdot {}^\circ\text{F}} \cdot (122 \, {}^\circ\text{F} - 104 \, {}^\circ\text{F})$

$Q_{\text{ammonia}} = 594{,}594 \, \frac{\text{Btu}}{\text{hr}}$

$Q_w = 501 \cdot \text{GPM} \cdot \Delta T$

$T_o = \dfrac{594{,}594 \text{ Btu}/\text{hr}}{501 \cdot 60 \text{ GPM}} + 60 \, {}^\circ\text{F}$

$T_o = 79.8 \, {}^\circ\text{F}$

The correct answer is B.

73. From ASHRAE HVAC Systems and Equipment, Chapter 46, the correct answer is A.

74. From ASHRAE Psychrometric Chart #1 at 80 °F$_{db}$ / 30% RH, $T_{\text{air, wb}} = 60$ °F$_{wb}$

$\eta_w = \dfrac{T_{\text{water,in}} - T_{\text{water,out}}}{T_{\text{water,in}} - T_{\text{air,wb,in}}}$

$T_{\text{water, out}} = T_{\text{water,in}} - \eta_w \cdot (T_{\text{water,in}} - T_{\text{air,wb,in}})$

$T_{\text{water, out}} = 95 \, {}^\circ\text{F} - 0.65 \cdot (95 \, {}^\circ\text{F} - 60 \, {}^\circ\text{F})$

$T_{\text{water, out}} = 72.25 \, {}^\circ\text{F}$

$Q = 501 \cdot \text{GPM} \cdot \Delta T$

$Q = 501 \cdot 600 \text{ GPM} \cdot (95 \text{ °F} - 72.25 \text{ °F})$

$Q = 6{,}838{,}650 \dfrac{\text{Btu}}{\text{hr}} \cdot \dfrac{1 \text{ Ton}}{12{,}000 \text{ Btu}/\text{hr}}$

$Q = 570 \text{ Tons}$

The correct answer is C.

75. From ASHRAE Psychrometric Chart #4,

$h_i = 33.0 \dfrac{\text{Btu}}{\text{lb}}$

$h_o = 43.0 \dfrac{\text{Btu}}{\text{lb}}$

From ASHRAE Fundamentals, Chapter 1,

$\rho_{\text{air, 5,000 ft}} = \rho_{\text{air, sea level}} \cdot \dfrac{\text{atmospheric pressure @ 5,000 ft}}{\text{atmospheric pressure @ sea level}}$

$\rho_{\text{air, 5,000 ft}} = 0.075 \dfrac{\text{lb}}{\text{ft}^3} \cdot \dfrac{12.23 \text{ psia}}{14.70 \text{ psia}}$

$\rho_{\text{air, 5,000 ft}} = 0.0624 \dfrac{\text{lb}}{\text{ft}^3}$

$Q = \rho \cdot 60 \dfrac{\text{min}}{\text{hr}} \cdot \text{CFM} \cdot \Delta h$

$Q = 0.0624 \dfrac{\text{lb}}{\text{ft}^3} \cdot 60 \dfrac{\text{min}}{\text{hr}} \cdot 15{,}000 \text{ CFM} \cdot (43.0 \dfrac{\text{Btu}}{\text{lb}} - 33.0 \dfrac{\text{Btu}}{\text{lb}})$

$Q = 561{,}581 \dfrac{\text{Btu}}{\text{hr}}$

$Q = 501 \cdot \text{GPM} \cdot \Delta T$

$\text{GPM} = \dfrac{Q}{501 \cdot \Delta T}$

$\text{GPM} = \dfrac{561{,}600 \text{ Btu}/\text{hr}}{501 \cdot 12 \text{ °F}}$

$\text{GPM} = 93.4 \text{ GPM}$

The correct answer is B.

76. From ASHRAE HVAC Systems and Equipment, Chapter 33, the correct answer is B.

77. $Q = U \cdot A \cdot \Delta T$

$\dfrac{Q}{A} = U \cdot \Delta T$

$\dfrac{Q}{A} = \dfrac{\Delta T}{R_{\text{outdoor air film}} + R_{\text{indoor air film}} + R_{\text{window}}}$

$\dfrac{Q}{A} = \dfrac{\Delta T}{\dfrac{1}{h_{\text{outdoor air film}}} + \dfrac{1}{h_{\text{indoor air film}}} + \dfrac{L}{k_{\text{window}}}}$

The correct answer is B.

78. $W = \dfrac{m \cdot a}{g_c}$

$W = \dfrac{2{,}600 \text{ lb}_m \cdot 32.2 \text{ ft}/s^2}{32.2 \text{ ft}/s^2}$

$W = 2{,}600 \text{ lb}_f$

$\delta = \dfrac{W}{K}$

$\delta = \dfrac{2{,}600 \text{ lb}_f}{4 \cdot 830 \text{ lb}_f/\text{in}}$

$\delta = 0.78 \text{ in}$

The correct answer is C.

79. $Q_{\text{air}} = 1.08 \cdot \text{CFM} \cdot \Delta T$

$Q_{\text{air}} = 1.08 \cdot 10{,}000 \text{ CFM} \cdot (160 \text{ °F} - 140 \text{ °F})$

$Q_{\text{air}} = 216{,}000 \dfrac{\text{Btu}}{\text{hr}}$

$Q_{\text{liquid}} = \dot{m} \cdot c_p \cdot \Delta T$

$Q_{\text{liquid}} = \dfrac{55 \text{ Gal}}{\text{min}} \cdot \dfrac{60 \text{ min}}{\text{hr}} \cdot \dfrac{1 \text{ ft}^3}{7.48 \text{ Gal}} \cdot \dfrac{62.4 \text{ lb}}{1 \text{ ft}^3} \cdot \dfrac{1.07 \text{ lb mixture}}{1 \text{ lb water}} \cdot 0.87 \dfrac{\text{Btu}}{\text{lb} \cdot \text{F}} \cdot (110 \text{ °F} - 102 \text{ °F})$

$Q_{\text{liquid}} = 205{,}017 \dfrac{\text{Btu}}{\text{hr}}$

$$Q_{air} = 1.08 \cdot CFM \cdot \Delta T$$

$$T_{out} = \frac{Q_{min}}{1.08 \cdot CFM} + T_{in}$$

$$T_{out} = \frac{205{,}017 \text{ Btu/hr}}{1.08 \cdot 12{,}000 \text{ CFM}} + 60 \text{ °F}$$

$$T_{out} = 75.8 \text{ °F}$$

The correct answer is B.

80. $f_{co} = \frac{c}{2 \cdot a}$

At 70 °F, $c = 1{,}130 \frac{ft}{s}$

$a = 20 \text{ in} \cdot \frac{1 \text{ ft}}{12 \text{ in}}$

$a = 1.67 \text{ ft}$

$f_{co} = \frac{1{,}130 \text{ ft/s}}{2 \cdot 1.67 \text{ ft}}$

$f_{co} = 338 \text{ Hz}$

The correct answer is C.

81. From ASHRAE HVAC Systems and Equipment, Chapter 47, the correct answer is D.

82. TEL = 1,300 ft + 650 ft

TEL = 1,950 ft

$f_{100} = \frac{ASP \cdot 100}{TEL}$

$f_{100} = \frac{1.6\text{" w.g.} \cdot 100}{1{,}950 \text{ ft}}$

$f_{100} = 0.082 \frac{\text{in w.g.}}{100 \text{ ft}}$

The correct answer is D.

83. For a single pass, counterflow heat exchanger,

$\Delta T_{approach} = T_{hot, in} - T_{cold, out}$

$\Delta T_{approach} = 110\ °F - 60\ °F$

$\Delta T_{approach} = 50\ °F$

The correct answer is C.

84. $A = \dfrac{16\ in \cdot 16\ in}{144\ in^2/ft^2}$

$A = 1.78\ ft^2$

$V = \dfrac{Q}{A}$

$V = \dfrac{6{,}000\ ft^3/min}{1.78\ ft^2}$

$V = 3{,}375\ FPM$

$VP = \left(\dfrac{FPM}{4005}\right)^2$

$VP = \left(\dfrac{3{,}375\ FPM}{4005}\right)^2$

$VP = 0.71\ in\ w.g.$

$TP = SP + VP$

$TP = 2.71\ in\ w.g.$

$AHP = \dfrac{Q \cdot TP}{6{,}356}$

$AHP = \dfrac{6{,}000\ CFM \cdot 2.71\ in\ w.g.}{6{,}356}$

$AHP = 2.56\ HP$

$ME = \dfrac{AHP}{BHP}$

$ME = \dfrac{2.56\ HP}{4.13\ HP}$

ME = 0.62

The correct answer is C.

85. From ASHRAE Psychrometric Chart #1, $T_{wb,\text{air in}} = 67.5\,°F$.

$$\eta_{sat} = \frac{T_{db,\text{air in}} - T_{db,\text{air out}}}{T_{db,\text{air in}} - T_{wb,\text{air in}}}$$

$$\eta_{sat} = \frac{80\,°F - 68\,°F}{80\,°F - 67.5\,°F}$$

$$\eta_{sat} = 0.64$$

The correct answer is D.

86. From ASHRAE Fundamentals, Chapter 7, the correct answer is B.

87. From ASHRAE Psychrometric Chart #1,

$$\rho = \frac{1}{v}$$

$$\rho = \frac{1}{14.05\,\text{ft}^3/\text{lb}_a}$$

$$\rho = 0.071\,\frac{\text{lb}}{\text{ft}^3}$$

$$\omega_1 = 0.009\,\frac{\text{lb}_w}{\text{lb}_a}$$

$$\omega_2 = 0.0215\,\frac{\text{lb}_w}{\text{lb}_a}$$

$$\dot{m}_{evap} = \rho \cdot \text{CFM} \cdot \Delta\omega$$

$$\dot{m}_{evap} = 0.071\,\frac{\text{lb}_a}{\text{ft}^3} \cdot 140{,}000\,\frac{\text{ft}^3}{\text{min}} \cdot \left(0.0215\,\frac{\text{lb}_w}{\text{lb}_a} - 0.009\,\frac{\text{lb}_w}{\text{lb}_a}\right)$$

$$\dot{m}_{evap} = 124.25\,\frac{\text{lb}_w}{\text{min}} \cdot \frac{1\,\text{ft}^3}{62.4\,\text{lb}} \cdot 7.48\,\frac{\text{gal}}{\text{ft}^3}$$

$$\dot{m}_{evap} = 14.9\,\text{GPM}$$

$$\dot{m}_{drift} = Q \cdot \% \text{ Drift Loss}$$

$$\dot{m}_{drift} = 550 \text{ GPM} \cdot 0.002$$

$$\dot{m}_{drift} = 1.1 \text{ GPM}$$

$$\dot{m}_{blowdown} = \frac{\dot{m}_{evap} + (1-C) \cdot \dot{m}_{drift}}{C-1}$$

$$\dot{m}_{blowdown} = \frac{14.9 \text{ GPM} + (1-5) \cdot 1.1 \text{ GPM}}{5-1}$$

$$\dot{m}_{blowdown} = 2.6 \text{ GPM}$$

The correct answer is A.

88. $R_{\text{outdoor air film}} = 0.17 \frac{\text{hr} \cdot \text{ft}^2 \cdot °F}{\text{Btu}}$

$R_{\text{siding}} = 0.15 \frac{\text{hr} \cdot \text{ft}^2 \cdot °F}{\text{Btu}}$

$R_{\text{cavity, avg}} = A_{\text{stud}} \cdot R_{\text{stud}} + A_{\text{insul}} \cdot R_{\text{insul}}$

$R_{\text{cavity, avg}} = A_{\text{stud}} \cdot \frac{\Delta x_{\text{stud}}}{k_{\text{stud}}} + A_{\text{insul}} \cdot R_{\text{insul}}$

$R_{\text{cavity, avg}} = 0.10 \cdot \dfrac{3.5 \text{ in}}{0.9 \,\text{Btu} \cdot \text{in} / \text{hr} \cdot \text{ft}^2 \cdot °F} + 0.90 \cdot 30 \frac{\text{hr} \cdot \text{ft}^2 \cdot °F}{\text{Btu}}$

$R_{\text{cavity, avg}} = 27.4 \frac{\text{hr} \cdot \text{ft}^2 \cdot °F}{\text{Btu}}$

$R_{\text{drywall}} = \dfrac{0.625 \text{ in}}{1.0 \,\text{Btu} \cdot \text{in} / \text{hr} \cdot \text{ft}^2 \cdot °F}$

$R_{\text{drywall}} = 0.625 \frac{\text{hr} \cdot \text{ft}^2 \cdot °F}{\text{Btu}}$

$R_{\text{indoor air film}} = 0.68 \frac{\text{hr} \cdot \text{ft}^2 \cdot °F}{\text{Btu}}$

$R_{\text{Total}} = R_{\text{outdoor air film}} + R_{\text{siding}} + R_{\text{insulation/ framing avg}} + R_{\text{drywall}} + R_{\text{indoor air film}}$

$R_{\text{Total}} = 29.0 \frac{\text{hr} \cdot \text{ft}^2 \cdot °F}{\text{Btu}}$

The correct answer is C.

89. $p_{steam, psia} = 20 \text{ psig} + 14.7 \text{ psi}$

$p_{steam, psia} = 34.7 \text{ psia}$

Interpolating for superheated steam tables,

$h_{steam} = 1{,}181.9 \dfrac{\text{Btu}}{\text{lb}_m} + (34.7 \text{ psia} - 60 \text{ psia}) \cdot \left(\dfrac{1{,}191.6 \text{ Btu}/\text{lb}_m - 1{,}181.9 \text{ Btu}/\text{lb}_m}{34.7 \text{ psia} - 60 \text{ psia}} \right)$

$h_{steam} = 1{,}188.0 \dfrac{\text{Btu}}{\text{lb}_m}$

From saturated steam tables,

$h_{feedwater} = 168.1 \dfrac{\text{Btu}}{\text{lb}_m}$

$Q_{in} = 135 \text{ kW} \cdot 3{,}412.1 \dfrac{\text{Btu}/\text{hr}}{\text{kW}}$

$Q_{in} = 460{,}634 \text{ Btu}/\text{hr}$

$\eta_{boiler} = \dfrac{\dot{m}_{steam} \cdot (h_{steam} - h_{feedwater})}{Q_{in}}$

$\eta_{boiler} = \dfrac{400 \text{ lb}_m/\text{hr} \cdot (1{,}188.0 \text{ Btu}/\text{lb}_m - 168.1 \text{ Btu}/\text{lb}_m)}{460{,}634 \text{ Btu}/\text{hr}}$

$\eta_{boiler} = 0.89$

The correct answer is D.

90. From ASHRAE Standard 34, the Refrigeration Concentration Limit for R-410A is $26 \dfrac{\text{lb}}{\text{Mcf}}$.

Max Charge = RCL · V_{min}

Max Charge = $26 \dfrac{\text{lb}}{\text{Mcf}} \cdot 4{,}800 \text{ ft}^3 \cdot \dfrac{\text{Mcf}}{1{,}000 \text{ ft}^3}$

Max Charge = 124.8 lbs

The correct answer is D.

91. The correct answer is B.

92. From ASHRAE Fundamentals, Chapter 15,

    $E_D = E_{DN} \cdot \cos(\theta)$

    $E_D = 1,400 \frac{Btu}{hr \cdot ft^2} \cdot \cos(90° - 30°)$

    $E_D = 700 \frac{Btu}{hr \cdot ft^2}$

    $q_b = E_D \cdot SHGC$

    $q_b = 700 \frac{Btu}{hr \cdot ft^2} \cdot 0.8$

    $q_b = 560 \frac{Btu}{hr \cdot ft^2}$

    $Q = q_b \cdot A$

    $Q = 560 \frac{Btu}{hr \cdot ft^2} \cdot 120 \text{ ft}^2$

    $Q = 67,200 \frac{Btu}{hr}$

    The correct answer is A.

93. From ASHRAE Fundamentals, Chapter 39, the correct answer is A.

94. From ASHRAE HVAC Applications, Chapter 42, the correct answer is A.

95. Space Effect = 25 − 10 log (ft) − 5 log (ft³) − 3 log (Hz)

    $SE_{125 Hz}$ = 25 − 10 log (18 ft) − 5 log (4,000 ft³) − 3 log (125 Hz)

    $SE_{125 Hz}$ = − 11.9 db

    $SE_{250 Hz}$ = 25 − 10 log (18 ft) − 5 log (4,000 ft³) − 3 log (250 Hz)

    $SE_{250 Hz}$ = − 12.8 db

    $SE_{500 Hz}$ = 25 − 10 log (18 ft) − 5 log (4,000 ft³) − 3 log (500 Hz)

$SE_{500\ Hz} = -13.7\ db$

$SE_{1,000\ Hz} = -14.6\ db$

$SE_{2,000\ Hz} = -15.5\ db$

$SE_{4,000\ Hz} = -16.4\ db$

$SE_{8,000,\ Hz} = -17.3\ db$

LP = Measured Sound Power Level + Space Effect

$LP_{125\ Hz} = 35\ db - 11.9\ db = 23.1\ db$

$LP_{250\ Hz} = 51\ db - 12.8\ db = 38.2\ db$

$LP_{500\ Hz} = 60\ db - 13.7\ db = 46.3\ db$

$LP_{1,000\ Hz} = 62\ db - 14.6\ db = 47.4\ db$

$LP_{2,000\ Hz} = 52\ db - 15.5\ db = 36.5\ db$

$LP_{4,000\ Hz} = 42\ db - 16.4\ db = 25.6\ db$

$LP_{8,000,\ Hz} = 45\ db - 17.3\ db = 27.7\ db$

Plot points on Noise Criteria Curve

Maximum Noise Criteria Rating is NC-40 at 1,000 Hz.

The correct answer is C.

96. $P_s = \dfrac{Q_s \cdot \Delta P_s}{6{,}356 \cdot \eta_f}$

$P_s = \dfrac{600 \text{ CFM} \cdot 1.0 \text{ in water}}{6{,}356 \cdot 0.7}$

$P_s = 0.13 \text{ HP}$

$P_e = \dfrac{Q_e \cdot \Delta P_e}{6{,}356 \cdot \eta_f}$

$P_e = \dfrac{500 \text{ CFM} \cdot 1.0 \text{ in water}}{6{,}356 \cdot 0.7}$

$P_e = 0.11 \text{ HP}$

$P_T = P_s + P_e$

$P_T = 0.13 \text{ HP} + 0.11 \text{ HP}$

$P_T = 0.24 \text{ HP}$

The correct answer is C.

97. From ASHRAE Psychrometric Chart #1,

$\nu_s = 12.7 \dfrac{\text{ft}^3}{\text{lb}}$

$\dot{m}_s = \dfrac{V}{\nu}$

$\dot{m}_s = \dfrac{4{,}000 \, \text{ft}^3/\text{min}}{12.7 \, \text{ft}^3/\text{lb}}$

$\dot{m}_s = 315 \dfrac{\text{lb}}{\text{min}}$

$\nu_e = 13.9 \dfrac{\text{ft}^3}{\text{lb}}$

$\dot{m}_e = \dfrac{4{,}000 \, \text{ft}^3/\text{min}}{13.9 \, \text{ft}^3/\text{lb}}$

$\dot{m}_e = 287.8 \dfrac{\text{lb}}{\text{min}}$

$q_s = 60 \cdot \varepsilon_s \cdot m_{\min} \cdot c_p \cdot (t_1 - t_3)$

$$q_s = 60 \cdot 0.52 \cdot 287.8 \, \frac{lb}{min} \cdot 0.24 \, \frac{Btu}{lb \cdot °F} \cdot (85 \, °F - 45 \, °F)$$

$$q_s = 86{,}200 \, \frac{Btu}{hr}$$

The correct answer is B.

98. Find the intersection of 6,000 CFM and the system operating curve. Drawing a straight line up until the next BHP line is reached yields 10 HP. The correct answer C.

99. From ASHRAE Fundamentals, Chapter 26, the correct answer is C.

100 $Q = \rho_a \cdot c_p \cdot V \cdot n_{AC} \cdot (T_i - T_o)$

$$Q = 0.075 \, \frac{lb}{ft^3} \cdot 0.24 \, \frac{Btu}{lb \cdot °F} \cdot 9{,}200 \, ft^3 \cdot 12 \, \frac{AC}{hr} \cdot (75 \, °F - 35 \, °F)$$

$$Q = 79{,}488 \, \frac{Btu}{hr}$$

The correct answer is D.

REFERENCES

1. ASHRAE Handbook: Fundamentals. American Society of Heating, Refrigerating and Air-Conditioning Engineers, 2013.
2. ASHRAE Handbook: Refrigeration. American Society of Heating, Refrigerating and Air-Conditioning Engineers, 2014.
3. ASHRAE Handbook: HVAC Applications. American Society of Heating, Refrigerating and Air-Conditioning Engineers, 2015.
4. ASHRAE Handbook: HVAC Systems and Equipment. American Society of Heating, Refrigerating and Air-Conditioning Engineers, 2016.
5. ANSI/ASHRAE Standard 62.1-2016: Ventilation for Acceptable Indoor Air Quality. American Society of Heating, Refrigerating and Air-Conditioning Engineers, 2016.
6. ANSI/ASHRAE Standard 15-2013: Safety Standard for Refrigeration Systems. American Society of Heating, Refrigerating and Air-Conditioning Engineers, 2013.
7. ANSI/ASHRAE Standard 34-2013: Designation and Classification of Refrigerants. American Society of Heating, Refrigerating and Air-Conditioning Engineers, 2013.

Made in the USA
Las Vegas, NV
01 December 2021